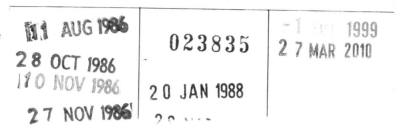

THE SPORTING GUN

THE SPORTING GUN

JAMES DOUGLAS

Line illustrations by

Ian Oates

0238351799·213

David & Charles

Newton Abbot   London   North Pomfret (Vt)

The author and publishers are
indebted to *Sporting Gun* magazine,
EMAP Publications, for giving their
permission for the title *The Sporting Gun*.

Line illustrations by Ian Oates
Line diagrams by Roger Thomas

British Library Cataloguing in Publication Data
Douglas, James
  The sporting gun.
  1. Shooting
  I.   Title
  799.2'13   GV1153
  ISBN 0–7153–8324–8

First published 1983
Second impression 1985

Photoset by ABM Typographics Ltd Hull
and printed in Great Britain
by Redwood Burn Ltd, Trowbridge, Wilts.
for David & Charles (Publishers) Limited
Brunel House   Newton Abbot   Devon

Published in the United States of America
by David & Charles Inc
North Pomfret   Vermont 05053   USA

Cock capercaillie

CONTENTS

To Shena, Morven and Isla —
who make it all worthwhile

Red deer hind

Partridge

INTRODUCTION

In this book I have tried to give the reader a comprehensive guide to all aspects of 'live' shooting and game lore. I have sought to provide in one book the answers to so many questions, from the correct choice of equipment and how to use it to how to find and identify a given species, ranging from a snipe to a stag. How correctly to go about the pursuit of a particular animal or bird, whether with gun, rifle, or camera, how to prepare the game for the table, using a variety of interesting, yet simple recipes, and the best methods for the care and display of trophies are tackled along with all the most pertinent aspects of fieldcraft that the complete modern shooter and countryman needs to know.

There is no mystique about the ability to read the myriad signs of the countryside. It takes only practice and patience, coupled with a good practical guide. Yet these skills, once learned, greatly enhance any trip outdoors, whether you are taking a stroll through an urban park, dawn flighting on a lonely, windswept foreshore, rough shooting on golden autumn stubbles, or up in the Highlands stalking red deer.

As the sportsman develops proficiency outdoors, as he becomes experienced in the ways of game and skilful at taking it, progressively his general appreciation of ecology, and particularly our wild creatures, changes. As the emphasis on killing is reduced, the true outdoorsman seeks more understanding and contact with nature. While he would still enjoy a day shooting, every aspect of the day gains greater importance, and I hope this book will give the fieldsportsman greater insight and understanding, and answer his questions.

Rabbit

1 THE SHOTGUN

At its simplest a shotgun is an engineering product lacking refinement or style and costing little; at its most sophisticated the culmination of centuries of development by craftsmen and artists whose finest expression is the classic side-by-side game gun sometimes costing thousands of pounds but worth every penny.

The choice of a gun may bewilder the newcomer but can be narrowed down by a process of elimination. Today guns in Britain are almost all imported from the principal manufacturing centres of Spain, Italy, the USA and Japan, and by and large fall into four neat categories according to their country of origin.

The American gun, usually self loading, commonly has a single barrel and magazine for up to five cartridges. This has been and remains the Americans' favourite tool, and for very good reasons. It is cheap, generally efficient and well suited to the pursuit of wildfowl or game birds by one man and his dog out for a days 'hunting'. The extra magazine capacity may be an advantage to the lone American sportsman but in Britain the use of such a gun in the company of others is rightly regarded as a potential danger — you cannot tell if it is unloaded — and it is thoroughly unsporting. Most hosts to a shooting party will not welcome such a machine on their land.

There are, however, some occasions when an auto-loading gun comes into its own. The wildfowler who does not relish the pounding of 3in magnum cartridges which he may use against geese or ducks on the foreshore will welcome the cushioning of the recoil which is a feature of such guns and it does not hurt so much to watch the effects of salt water and mud on a gun which will have cost probably a quarter of the price of a modest secondhand English side-by-side boxlock.

Most of us in Britain would choose to shoot game with an English gun and that means a lightweight, side-by-side double-barrelled boxlock or sidelock. The history of the English sporting gun has been one of refinement and elimination of every aspect of the product not absolutely essential to the total enjoyment of the sport. Weight is reduced to the minimum consistent with sufficient strength to stand up to continuous use. All the functions are simple, easily maintained in good working order and, if the gun is correctly fitted, it will per-form as effortlessly as a violin in the hands of a musician. To achieve such simplicity requires the best materials and calls for the skills of craftsmen whose talents do not come cheaply. Sadly, therefore, the English gun, produced in the traditional way from finest steels, stocked in walnut nowadays mostly only used for veneering, and painstakingly hand-engraved, is within the financial reach of very few. It has become an investment item and frequently never sees use in the field. The bank vault is often where it stays while the owner, like most of us, prudently avails himself of the more accessible pro-ducts of the Spanish gunmakers.

Spain now supplies the vast majority of the traditional type of side-by-side double-barrelled guns normally used in the British sporting field. As in any competitive industry the quality of Spanish guns ranges in more or less direct ratio to their price. The finest products of today's Spanish gunmakers are superlative in all respects while at the other extreme the guns may lack many of the refinements of their more expensive fellows but at least are sure to be safe.

In Spain, as in Britain, France, Germany, Belgium, Austria and Italy, all guns must, by law, be tested at an independent Government Proof House under strict supervision, to ensure that they will stand up to the highest pressures developed by the cartridges for which the gun is designed. A word of warning — the length of chamber of each gun is stamped on the barrels after proof — and that applies to all guns sold in Britain, even if they come from countries which do not have their own official Proof House, in which case they must, before they are sold, be tested and marked at one of the two British Proof Houses in London or Birmingham — and you should only use car-tridges of a length which corresponds to, or is shorter than that indi-cated on the gun.

It is easy enough to distinguish a cheap gun from an expensive one by a glance at the outside. The fit of wood to metal, the finish, the quality of the walnut, the smoothness and ease (or lack of it) with which it opens and shuts, crisp, neatly-cut chequering (the crisscross lines cut into the grip and fore-end to provide a firmer hold) and finely executed engraving are all hallmarks of a gun's quality; but a boxlock or a sidelock? Ejector or non-ejector? The differences add up to many pounds and may seem hard to justify. The answer is to buy the most expensive gun you can afford and it will be its own reward. Many a sportsman on a January morning has rued the day he bought a non-ejector gun as he struggles with numb fingers to pull fired cases out of the chambers while the only flush of birds that day clatters over his head. Another £50 or so would have made reloading so much easier and quicker.

A boxlock does just the same job as a sidelock. The difference is that the latter has two separate lock mechanisms, one for each barrel, and is therefore, theoretically at least, less likely to let you down completely if anything should go wrong with the works. The sidelock sacrifices a very little in weight on account of the extra lock but is closer to the tradition of best English guns and to most gun owners is aesthetically more pleasing than the humbler boxlock.

There is, of course, more to choosing a gun than deciding on your favourite colour, but before considering such matters as choke, barrel length or stock dimensions there is a whole category of shotgun not yet mentioned — the over-and-under.

A number of factors have contributed to the greatly increased popularity of the over-and-under (or under-and-over) in Britain. This is a gun with two barrels arranged vertically, one above the other as opposed to the side-by-side. The arrangement of the barrels allows the action body to wrap around and support them more securely than in the side-by-side where the design relies on lugs projecting from the underside of the barrels for their attachment to the action. There are some advantages and some drawbacks. The drawbacks are greater weight since there is more metal in the action and an extra rib on the barrels. To load the cartridges the barrels have to be opened much further than in a side-by-side to allow clearance for the bottom barrel above the top of the action. Some find the over-and-under clumsy in appearance because of its deep profile, but nevertheless to many its advantages outweigh these slight objections.

The advantages of an over-and-under are paradoxically those very features which in the best side-by-side are considered undesirable. This is because the guns are based on different concepts. Try shooting more than twenty or so cartridges from a lightweight side-by-side in quick succession and you will have a scorched hand and probably a sore shoulder. Those lucky enough regularly to shoot a large number of driven pheasant will welcome the sore shoulder and wear a glove or use a leather-covered spring-heel hand protector on the exposed barrels, but if you have to content yourself with a Sunday morning at the local clay pigeon club when it is not uncommon to fire 100 cartridges, the over-and-under comes into its own.

It is the growth in popularity of clay pigeon shooting which has led to much wider use of over-and-under guns. They are robust and heavy enough to absorb recoil, making prolonged shooting more comfortable. The fore-end wood wraps round the barrels shielding the hand from the heat. Looking down a single barrel makes sighting easier for many shooters and the fact that, unlike game shooting, clay pigeon shooting often involves deliberate preparation before calling for a bird, makes the superior ease of gun mounting (getting it quickly into the shoulder) of a side-by-side less important.

Almost all over-and-unders have a single trigger, usually of the selective type which enables the shooter to choose whether he fires the top barrel first or the bottom. The mechanism automatically switches to the second barrel after the first has been fired, either by a mechanical spring system or by utilising the inertia of the recoil from the first shot. Having only one trigger, the hand remains in the same position on the stock for both shots so making a pistol grip more comfortable than would be the case on a gun with two triggers where the hand is slid backwards to reach the second trigger between shots.

So many people have come into the sport through clay pigeon shooting that it is not surprising that, having become accustomed to the handling of an over-and-under, they use this type of gun for game and vermin shooting too. Add to this the very wide range of models in all price brackets now available on the market and you have the ingredients for a challenge to the supremacy of the side-by-side.

If Belgium and Germany were the birthplaces of the over-and-under, their present day homes are Italy and Japan from which countries the vast majority of these guns now issue. Modern engineering methods and a highly productive industry have resulted in these countries supplying a range of over-and-under guns which covers the whole spectrum from the cheap but serviceable to the height of sophistication in gunmaking.

Having decided on your preferred sport and shaken your piggy bank you have probably concluded that you will choose between a side-by-side or an over-and-under but then you have only just started. Some of the subsequent choices will be determined by the weight of the piggy, such as whether to have gold inlaid mallard on one sidelock and grouse on the other, or whether to carry your plain boxlock non-ejector in a plastic gun case or a canvas one. Other decisions have to be made first, however, on points such as barrel length and chokes, stock dimensions and rib type, single trigger or double and, of course, gauge or bore.

Most game, and practically all clay pigeons, are today shot with a 12-bore, but for a youngster to start shooting at 10 or 12 years of age with such a powerful cartridge may be impractical and there are lighter, smaller gauges to start on. The .410 is the smallest and not recommended for anything more than small ground game or vermin at short ranges, but a 20-bore is light on the shoulder and will kill effectively at ranges almost as great as a 12-bore. It is misleading, however, to talk of range with a shotgun since the pellets from a 12-bore are travelling at much the same velocity as those from a .410 and it is their greater number which makes the 12-bore suitable for shots at ranges of 30-40 yards — not more. No-one with a sense of responsibility would attempt long shots at live game beyond about 40 yards with a 12-bore as the shot at that range is usually so scattered and its energy so diminished that wounding is almost certain.

Shot patterns, the funnel-shaped spread of the pellets as they leave the muzzle, are determined mainly by the choke or constriction inside the barrel just before the muzzle. A full-choked barrel should concentrate 70 per cent of the pellets inside a 30in circle at 40 yards. This percentage will diminish through a succession of lesser choke constrictions until, if there is no choke or the barrel is 'true cylinder' there will be about 40 per cent of pellets in the circle. For shooting live game it is advisable to have one barrel lightly choked, say ¼ or improved cylinder and the other ½ or ¾ (modified) choke. If you have too much choke a bird shot at close range will be plucked and dressed before it hits the ground and only good for the dog. Alternatively too little choke will restrict you to short range shots or tempt you to leave a trail of wounded and crippled game through having attempted to shoot at ranges at which the pattern is spread too thinly to be lethal.

At this point it should be said that the cartridge may influence the pattern from a barrel almost as much as the choke. Cartridges loaded with a plastic, cup-shaped wad which contain the lead shot and keep it together during its passage up the barrel have a tendency to concen-

13

trate the pattern much more than those which have a felt or composition wad separating the shot from the powder. It it not advisable to use cartridges with the plastic wads, which are designed primarily for clay pigeon competitions, in the game field. If nothing else they will infuriate the farmer by littering his land with eternally offensive debris.

It is dangerous to use an old gun with modern cartridges unless it has been proved for nitro powder. There are hundreds of venerable fowling pieces about, many very fine and tempting to use, but even more in such a state of dilapidation or so lightly built that to fire them with modern high pressure cartridges would be like playing Russian roulette. Any gunsmith will be able to check both the proof marks and the fitness of a gun for use. It is also an offence to sell or exchange a gun which is out of proof — or outside the limits of safety as defined by the Gun Barrel Proof Act — so an old gun should be looked over by an expert gunsmith before it is either fired or sold.

Mass production methods which bring shotguns within the financial reach of the majority necessitate standardisation of stock dimensions which inevitably means that you will be lucky if a sporting gun bought off the peg fits you as perfectly as if it had been tailored to your size. If a stock is too short, the gun may feel comfortable in the shop but will probably clout you in the face and shoulder when fired. If it is too long it will be difficult to mount quickly, catching in clothing, particularly in winter, but this is a fault on the right side. It is easy enough to shorten a stock but lengthening one which is too short involves either fitting a rubber recoil pad or grafting on an extra lump of walnut which, though effective, can be unsightly and expensive since it involves refinishing the whole butt and often fitting a new and larger heel plate. If you can afford to do so, have a gunsmith measure you up using a try-gun — a gun with an adjustable, articulated stock with which not only length, but also the 'drop' of the comb, 'cast off' for right-handers or 'cast on' for left-handed shooters and so on can be accurately measured.

While it is always possible to adapt your shooting position to suit a gun which does not quite fit, it helps enormously in the game field to have one which comes up naturally into the correct position. Usually there is little time to prepare for a shot on the wing and an ill-fitting gun will not come up quickly or easily into the shoulder.

A well-made shotgun will last for many years and provide sport and enjoyment for your grandchildren if treated well. There is an

14

old saying that a new gun stock should be oiled like a cricket bat — treated with a light coating of linseed oil once a day for a week, once a week for a month, once a month for a year and once a year for the rest of your (or its) life. With synthetic varnish used more and more on gun stocks this saying has rather less relevance now than it used to, but regular cleaning and care in the storage of guns are of great importance.

The great enemy of all guns is rust. After a day out in torrential rain your gun will not appreciate a night in the boot of the car in a sodden cover. Kit yourself out with a good cleaning rod and implements of the right size — wool mop, jag or loop cleaner for cloth patches, phosphor-bronze wire brush for scrubbing out the accumulated powder residue and lead traces left by the shot and plenty of oil or solvent; solvent for cleaning the inside of the barrels and a lubricating and rust-preventing oil for the outside metal work. It is not a good idea to pour great quantities of cleaning solvent into the action as this will eventually solidify, gumming up the works which will then have to be completely stripped down and cleaned.

After shooting, the bores should always be scrubbed with a phosphor bronze wire brush and cleaning solvent, dried with cloth patches until they come out clean, then lightly oiled with a wool mop and rust-preventing oil. Wipe the outside of the gun with an oily cloth to remove all moisture, mud and powder residue and do not put it back into a wet gun cover or all the cleaning will have been in vain.

Cartridges

I am often surprised by the large number of people who hold erroneous views on cartridges and their capabilities. Yet, when questioned, they will confess never to have patterned their gun with a range of cartridges to give themselves an accurate picture of the cartridges' suitability for their particular gun and its choking. In addition, in recent years, with the increase in availability of European and American cartridges, there is a school of thought wanting more power, under the mistaken belief that somehow it gives the shooter an advantage. The following table will give you a rough guide on choice of shot size:

Target	Shot size
Geese	1 or 3
Hare	4 or 5
Mallard	4, 5 or 6
Rabbit	5 or 6
Pheasant	5, 6 or 7
Pigeon, Grouse Partridge, Teal, Woodcock, Squirrel	6 or 7
Snipe	7 or 8

However, the requirements of the shooting man can be narrowed down to no more than two or possibly three different loads. For the average game shooter, 5s or 6s will happily accommodate virtually every species of game he is likely to shoot in Britain on the small game list — the largest being a hare and the smallest being a snipe. If the shooter intends specifically to shoot snipe then 7s or 8s would be a better choice, but few men indeed shoot more than a handful of snipe in a season. As far as heavier birds such as geese are concerned, the best cartridges are 3s, and it is wrong to use BBs for geese, although there are some authorities who do recommend them. The reason I would never use BBs is that they only have 70 pellets in 1oz, as compared to 140 pellets with 3s. I have found the correct balance of density of pattern with weight of shot to be 3s, as I have stated, with a second choice of 1s which have 100 pellets to the ounce. The argument for BBs is normally a variation of the same theme, that it only requires one or two in the right place, and is a ludicrous statement. Striking the bulk of the bird's body area with a few pellets will not prove immediately fatal and will allow it to fly on. The aim should be to strike the birds with a dense pattern of pellets, greatly increasing the chances of striking vital areas. In addition, if you are swinging well enough in front when shooting birds within range, these heavy shot sizes are completely unjustified.

Most modern 12-bore shotguns are chambered either for the $2\frac{1}{2}$in (65mm) or $2\frac{3}{4}$in (70mm) cartridge. You may find this confusing if you measure the cartridge case itself as often it will not measure exactly 2½in or 2¾in. The designation refers more to the pressure developed by a certain cartridge than to its outside dimensions, so be sure that the chamber length printed on every box of cartridges is not longer than that for which your gun has been proof tested. Watch out also for 'magnum' cartridges which come in 2¾in and 3in (75mm) chamber lengths and which require a gun capable of standing up to much higher pressures than a standard 2¾in chamber.

As far as the load is concerned there is a school of thought that favours heavier loads — Hymax, Alphamax and magnums. Again, the argument for their use is an extremely thin one. For a fractionally faster speed and a fractionally denser pattern the shooter gets a great deal more recoil, has to pay higher prices, and risks damaging his gun with the heavier pounding. It is just not worth it. The whole secret of good game shooting is, as I have said, that you correctly pattern your gun, and this must be done with a variety of cartridges. How to do it is very simple. On a large sheet of plain white paper draw a .30in circle. At 40 yards shoot at the circle. The bulk of your shot should be contained within the circle, with a nice overall pattern. You will be surprised with the variations you can get between one cartridge load or manufacturer and another. All you must do is pick the one which gives you the best pattern.

Shotgun Use

Whole books have been written on how to use a shotgun, invariably going into numerous unnecessary bywaters, dazzling the reader with confusing and useless facts. However, there are a number of solid ground rules which must be adhered to if you wish to be a good, productive, safe shot. Mention cannot be made too often of the ever-present need for shotgun safety. Every season people either shoot themselves or are shot by their companions, yet all these dreadful accidents could so easily be avoided if that little extra care was observed by everyone. The two main categories of individuals who have accidents are the rank beginners and, probably more commonly, the experienced men who have become careless, allowing their familiarity with shotguns to give them a false sense of security.

The most common accidents seem quite incredible when one hears the facts; the loaded shotgun that goes off in the Land-Rover, almost cutting the occupant in half; two friends sharing a hide, one stands up to shoot while the other shoots from a sitting position, blowing his friend's head off; the man with the closed gun tripping in a root field and shooting the next man down the line in the side; in the goose hide the precariously balanced loaded gun which is knocked over by the dog and shoots his master; and very frequently during drives guns peppering beaters and other guns as the sportsmen become so intent on the quarry they forget all semblance of sense.

These accidents are no invention, they all happened in the few months preceding the writing of this book, and all of us must be ever vigilant when using shotguns. We must never have a loaded shotgun in a vehicle, we must always break our guns and empty them before crossing a wall or fence. We must never get into the habit of pushing the safety off before we mount the gun; the safety catch should always be slid off as the gun is mounted, and when walking in line guns should be broken. If your friend or acquaintance at any time

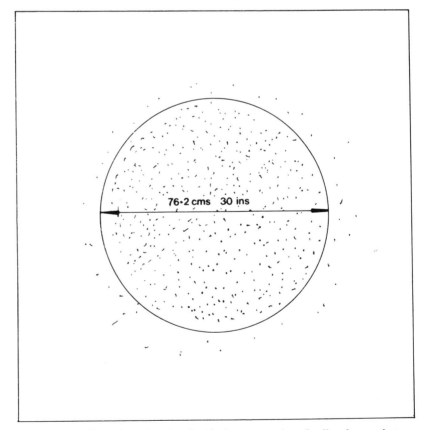

76·2 cms   30 ins

A 30in circle at 40 yards — showing the ideal concentration of pellets from a shotgun

points his gun in your direction, whether you know it to be unloaded or not you must immediately chastise him, making him aware of his actions and possibly preventing yet another nasty incident.

Shooting

All too often the reasons men miss with a shotgun can be put down to two — a lack of the correct attitude and preparedness, and the inability to judge distance and speed. Let's take driven shooting first. Whether you are standing in a root field or in a grouse butt, you must control the tendency of over excitement and keenness which results in you trying too hard. Most of us have had the experience of watching a bird coming straight toward us, from a fair distance away. Despite plenty of warning we swing up and miss. Invariably the reason for this is that you have mounted the gun too soon and ended up aiming it like a rifle, missing behind. Ideally of course you want to try for perfect timing — refrain from mounting until the bird is almost overhead by lifting the gun in one smooth action to your shoulder, swinging through and taking the bird when it is almost straight above you. To illustrate this point further, anyone who has flighted pigeons as they come in to roost, getting fleeting glimpses of the birds as they pass over short gaps between the treetops, normally finds that his kill ratio increases. The point is that you have not been given time to think about it, but swung through and shot on first aim. You have not been able to wonder if you are correct and start to change your point of aim.

The correct technique of game shooting should always take the following form. Break your gun letting the barrels fall. With the barrels pointing at the ground hold the gun in your left hand and drop in two cartridges. Hold the stock in your right hand where it feels comfortable, and with a slight lean forward lift the stock to close the gun. For left-handers the opposite hands would apply. It doesn't matter how the individual prefers to execute this movement; the only criterion is that the barrels must *always* be pointing at the ground.

When taking a standing shot mount the gun keeping both feet flat on the ground and throw your weight forward on to your left leg. When taking an on-coming overhead shot mount the gun leaning back with your weight on your right leg, and swing through, taking the bird just before he is straight above you. For sitting shots in a hide the best method is to sit with the gun between your knees with the

19

stock resting on the ground, holding the fore-end with your left hand. Again mount the gun at the last moment prior to shooting. Throw the gun up and as it seats into your shoulder swing sweetly and shoot.

The biggest problem people have when shooting from a hide, particularly with geese, which seem capable of engendering nerves in some of the coolest individuals, is to decide *before* you are going to shoot, particularly if a friend is sharing your hide, whether you intend to stand or sit when taking the shot. Normally, if you are sharing a hide, since the birds will approach the hide at different angles, you would do this as you see the bird approach. But whatever you do both you and your shooting companion must do the same.

If you are going to take a standing shot from sitting in a hide you must practise mounting the gun in one fluid motion. Throwing the gun up in front, you straighten your legs and come up behind it, ideally the stock connecting with your shoulder as your knees straighten. This is an easy technique to learn. Do not try standing and then lifting your gun if in a hide, since the barrels should be pointing up the way. If you stand up you will find the barrels invariably pointing straight up your nose. Equally, if you have your gun held as you would otherwise correctly hold it outside the hide, that is with barrels pointing down, in the heat of the moment as you try to lift your gun up to get a shot, you will discover that nine times out of ten the barrels will catch in your net.

You will find that it really pays dividends to adopt a cool, confident attitude and avoid over-anxiety. Keep your gun down, mounting it at the last minute, then swing and shoot. Two conflicting examples illustrate this point.

On the one hand I know an elderly man who shoots pigeons on a regular, semi-professional basis. He sits in his hide seemingly oblivious to the birds as they come in on his decoys. Then in the last few seconds he casually lifts his gun and shoots. He has done it so often that he never gets anxious or over-keen.

On the other hand I remember being asked to photograph a famous international field trials champion dog, and had to have some freshly shot cock pheasants to decorate the picture. Walking up a root field which had not been shot that year and that I knew to be bulging with birds, I missed time after time. My over-anxiety was playing havoc with my judgement. When I thought we had emptied the field of birds I stopped in disgust. As I was lighting a cigarette two birds were put up. I threw the gun up and had a neat right and left, as several more sprang up at the sound of my shot. Without stopping to think about it I spun, re-loaded, and had two more, neat as wink. It had happened so fast, catching me unawares, and I took each of the birds on my first aim.

The second main problem the majority of people suffer from is an inability to judge distance and range. The perfect place for judging distance is when using decoys and a hide. If you have decoys out you should know exactly how far the nearest and furthest are, since you would have paced out the distance from your hide. A good technique if you are slightly rusty on judging range is to place a few stones in a measured arc around your hide. These stones should be at your maximum range of 50 yards (pace it out).

The surprising fact is that most game is shot at distances between 15 and 20 yards, with the *effective* range of a shotgun being a little over 40 yards. Confusion arises for a number of reasons, not least the ridiculous claims made by some individuals as to how far their gun or specific cartridge will bring down game, and how they have shot exceedingly high birds with ease. A number of other contributory factors, such as the size of the individual species, can lead the individual to misjudge distance. The larger species of game, particularly hares and geese, seem closer because of their size. This results in far too many being shot at out of range, and missed, or worse still, being wounded.

Now while it is relatively easy to teach yourself how to judge distance across ground, a bird in the sky, with no reference points to aid you, is much more difficult. But think of it this way — the average very tall tree in Britain is unlikely to be more than about sixty-feet tall. Your maximum range therefore is likely to be about twice the height of any surrounding tall trees. If you can teach yourself how to gauge distance accurately, and you have patterned your gun with the relevant cartridges (see **Cartridges**) you should have few problems.

Greylag goose

2 GAME SHOOTING

Game shooting falls into several different categories, and for the newcomer to the sport, these can be confusing — formal shooting, syndicate shooting, rough shooting, and wildfowling.

Formal Shooting

On a formal shoot guests would normally arrive on the estate at the appointed hour, usually about 8 or 9am, and may have a light breakfast with the host before being introduced to the headkeeper. It is the headkeeper's day, when he will see his hard work of the previous year coming to fruition.

After introductions, in consultation with the host, the headkeeper informs the guns which numbered pegs or butts they have been allotted (sometimes the guns themselves draw for pegs). The individuals then make their way either by vehicle or foot to their stations, to wait quietly while the beaters start to drive the birds.

The major emphasis in formal shooting is on the difficulty of the shot, making the bird as sporting as possible. When shooting pheasants the guns would normally be placed in a ride, or outside a belt of trees, preferably on a slope where the birds can be presented to them at maximum speed and range. Depending on the scale of the shoot and the expectation of birds, some guests may take a loader to load the second of their pair of guns as they shoot with the first. After each drive the keepers and the pickers-up collect all fallen birds, prior to the line moving to the next position, where the procedure is repeated. Often a horn or whistle will be used to signify the start and end of shooting on a drive and must be strictly obeyed.

After lunch shooting may go on all afternoon prior to guests being offered duck shooting in the evening, if available.

There is a certain amount of etiquette which must be observed when you are invited to a formal shoot. It is not permissible to leave the numbered peg you have been allotted even if you think that by moving just 20 yards you could get better shooting. You must never take a bird which is within the sphere of the next gun in the line. Imagine an invisible line midway between you and the next peg and take birds only on your side of that line. In the event of you wishing to take a bird behind, which has passed through the line, you must break your swing, lift the gun vertically, and turn completely around before re-mounting the gun and making the shot.

If you have a dog it is not advisable to take it to a shoot unless it is well-trained and rock steady. If you are in the slightest doubt as to its steadiness when birds are tumbling around your ears, then it is advisable to peg it down when you arrive at each position, since, apart from saving yourself a vast amount of embarrassment, if your dog goes tally-ho after a runner, it may well be running through the next covert which has yet to be shot, to the keeper's dismay and your disgrace.

While formal shooting has many devotees, it also tends to have a considerable number of 'social' shooters — individuals who have taken up the sport either because it is socially acceptable or may benefit them professionally, and invariably it is with these people that accidents occur. Due to the enormous expense of mounting such an event — laying on birds, beaters, keepers etc — formal shooting is sadly becoming less popular as few people can afford the outlay.

At the end of the day it is always good manners to enquire of your host how much of a gratuity the guns should collect for the keeper. You will be expected to pay toward the cost of beaters, but if you have had a good day it is always advisable to make some contribution to the keeper, whether the host thinks it necessary or not. I am reminded of an amusing incident that took place some years ago on an estate where a friend of mine was keepering. Ron, a wily, droll character, was traditionally given various gratuities by the members of the shooting syndicate on the last grouse shoot at the end of the year. At the time it was usually £2 or £3, which in those days was quite a lot of cash. As the various guns came up and handed Ron a few folded notes, one extremely mean, though wealthy, retired brigadier discreetly handed Ron his donation, which Ron quietly dropped under his foot and then he proceeded to look extremely busy as he searched for something in the long grass. Eventually the other guests enquired of Ron what he was looking for. 'Oh', said Ron, 'The brigadier gave me half a crown as a tip, and I've dropped it.' The brigadier was dropped from the syndicate the following year, and probably never knew why. The moral is, if you have a good keeper, look after him, they do not grow on trees.

Syndicate Shooting

A form of shooting which is increasingly popular is one where a group of individuals have pooled their resources to create a small syndicate, doing the work themselves, leasing a piece of ground, working at clearing rides on weekends throughout the close season, improving ponds or wet areas, building hides, trapping vermin and sharing the day-to-day tasks which, on an estate, would be carried out by the keepers. From their collective budget these small do-it-yourself syndicates normally put down as many birds as they can afford. The ingenuity displayed by some of these syndicates is most laudable, often with members having fairly complex rearing programmes in their own gardens, where the birds can be carefully husbanded prior to being put in a release pen on the shoot. If the syndicate is run properly, with several organised days, it can be great fun and very rewarding, providing sport for men who could not otherwise afford it. It is most important that the members elect a shoot captain, whose responsibility it is to co-ordinate the efforts of the members and who is in charge of shooting days.

Although I have known of some very good syndicates to exist without a change of members for many years, it seems, sadly, that a number of them are doomed to failure due to personality differences, greed, or some members not being able to put in as much work as others. Invariably this leads to friction and disillusionment. Another hurdle which syndicates seem to face is that with growing costs there is the ever-present temptation to increase the membership, and problems seem to increase in direct proportion to the number of members. The golden rule, whether you are forming a syndicate or joining one, is that small is beautiful, and the number of members should be limited by two factors — the number of guns the ground will comfortably sustain, and how much you can afford.

Rough Shooting

Rough shooting is by far the most popular, in fact the mainstay, of British shooting sports. The term 'rough shooting' is all encompassing and at times misleading. The typical rough shooter has managed, either on his own or in the company of a few friends, to procure the shooting rights of a piece of land. Here he does his own keeping and husbandry of any game that may be there, controlling the vermin, and if he is keen and can afford it, putting down a few birds. Invariably he is the sort of man who is knowledgeable on and interested in the ways of the country and the habits of game. Usually he has a great appreciation of the game that he shoots, since he has to work so much harder for it and it is all destined for his own larder.

To my mind rough shooting can be the most pleasurable kind of game shooting, since the emphasis is normally on a day pottering about on your own ground with your dog, taking a mixed bag of whatever is on the ground — hares, pheasants, partridges, pigeons, ducks. The rough shooter is involved on a regular basis with what is on his ground, and can derive a great amount of pleasure from seeing his own work coming to fruition and making his own decision as to how much he wants to shoot and how often. If the game stocks permit it he may still have a few friends in for a driven day, but generally it is a lack of formality and being able to his own thing that attracts him. Normally he has his own dog which he has trained himself, and is more likely to have been attracted to the sport for the most genuine of reasons, and not simply as a social pastime, since rough shooters generally prefer to enjoy a fairly solitary pleasure.

Wildfowling

Wildfowling is really a variation of rough shooting, and most rough shooters do a bit of fowling if they have water on their ground. The term wildfowling is loosely used to cover the shooting of ducks and geese, though correctly it applies to foreshore shooting and those hardy individuals who shoot the marshes, estuaries, and sandbanks of the tidal areas around our coasts. Few people indulge in proper wildfowling, most being rough shooters who go fowling on occasions, though there are men whose sole shooting interest is in foreshore shooting.

If fowling attracts you it is imperative that you do your homework. You must satisfy yourself as to access, tide times and the weather report (from the Meteorological Office or coastguard) before shooting on any foreshore area, and you should carry a compass. Familiarise yourself with maps of the areas and tell someone where you intend to go and your expected time of return. Every shooting season many fowlers get into difficulties, though one hears only of the very small number of fatalities. Most incidents are avoidable if care is taken by the fowler.

Shooting in Britain has never been more difficult to come by than it is for the sportsman of the eighties. Gone are the days when virtually any farmer would happily give permission for the sportsman to wander at will, enjoying a day's rough shooting. As a youth the area where I had permission to shoot measured some ten miles by eight, and I could have had more, but never got around to asking. Such things are now of course virtually impossible to achieve without the aid of a very great deal of money. More are looking for shooting and, consequently, with greater demand at the cheaper end of the market — ie rough shooting, the whole sport has become cash orientated.

Also, with the general reduction of the rabbit population the sportsman has to make do with less game available. Hence the emphasis on conservation in recent years. The wise, modern rough shooter now cherishes and nurtures any land he is lucky enough to get access to, and a whole new appreciation has grown up of even the humble woodpigeon, with sportsmen discovering that this bird can offer exciting and challenging shooting.

This greater pressure on whatever shooting is available necessitates that every field sportsman and woman in Britain actively pursues a policy of improvement of both the ground and the species thereon. It makes no sense at all to wipe out the rabbits and, though this is opposite to the landowners' desires, a fine balance must be struck between each interested party.

Probably never in the history of British field sports has there been a greater incentive to the individual to create flight ponds out of any small, wet hole, to persuade the farmer to plant the sort of crop that is likely to encourage and hold game, and most important of all, to shoot sparingly, having at all times an eye permanently fixed on the future, on next year's breeding stock, and to attempt to live in harmony with all animals on the game list, taking only sufficient for his needs off the top of the population, and doing his utmost to ensure the success of following seasons.

Ground Game for the Shotgun

Rabbit *(Oryctolagus cuniculus)*

The rabbit has for many years been the mainstay of the British countryman, providing an extremely valuable source of food and fur,
though the fur has had less importance in this century. Surprisingly
the rabbit is not indigenous to this country, its origins being in
southern Europe and North Africa. It was probably introduced to
Britain by the Normans as a ready supply of good, highly nutritious
and fresh meat.

Originally rabbits were kept as domestic animals prior to their
escape and establishment in the wild. These wild rabbits appear to
have remained fairly localised until the mid-1800s since most of them
were kept in enclosed or walled areas where warrens had been established. However, once they escaped from these enclosed areas their
number multiplied and leap-frogged as they established themselves
and eventually they covered the whole country. Although in times of
national necessity, such as during the two World Wars, rabbits provided a most welcome and much valued meat source for the population, as agriculture became far more intensive it highlighted the
enormous threat from and damage to crops by the vast population of
rabbits.

During the mid-1950s the disease myxomatosis was introduced
and decimated the rabbit population throughout the country, and for
several years it became a fairly rare animal. Since the first epidemic of
this disease which virtually wiped them out, leaving only small
pockets, rabbits have never regained their former numbers. However, in some localities rabbits have started to multiply and seem to
have developed a degree of resistance to the disease, though as soon
as their numbers build up and a period of damp weather occurs,
when the rabbits are likely to huddle together underground, the conditions where the disease can flare up and reduce the population once
again are apparently created.

Rabbits are polygamous, one buck normally serving several does.
The gestation period is 28 days and litters may number anything
from three to nine in a single year, the larger litters being born during
the summer months. Rabbits are helpless for the first three weeks of
their lives but the doe makes an excellent mother and will defend her
litter with ferocity.

If you are fortunate enough to have rabbits on your ground you

27

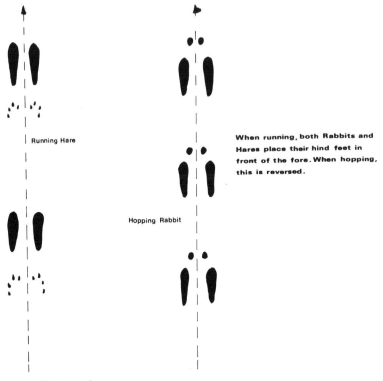

Running Hare

Hopping Rabbit

When running, both Rabbits and Hares place their hind feet in front of the fore. When hopping, this is reversed.

Rabbit and hare tracks

must use your intelligence. Farming interests will want them eradicated but you would be extremely foolish to wipe out such an excellent source of food and sport; it is better to reach a compromise and shoot them fairly vigorously, taking care never to persecute the population too efficiently. If the farming community insists on sterner measures, then ferreting, either by yourself or using someone more experienced, is by far the best method. At all costs you must try to avoid the rabbit population being gassed, since once a farmer or landowner embarks on this practice he seems incapable of stopping until the whole population has been eradicated, with obvious detrimental results to the shooting interest.

Rabbits are vegetarian, eating a wide variety of shoots and grasses, but will very occasionally eat an earthworm or snail. Many herbivores follow the practice of eating their droppings in order to extract any nutriments they may have missed in the first digestion. This is

28

done only once. On the second occasion the droppings will be left. The smell tells the animal whether the dropping has been digested once and should be eaten, or digested twice and left. Rabbits are no exception in following this practice, known as refection. When the droppings are to be left they are deposited in a latrine or dropping area, often on mounds or an old ant hill. These are easily spotted wherever rabbits live.

One result of the decimation of the rabbit population in some areas was the reduction in prey for predators. The decline of this food source was reflected in a number of changes. The majority of predatory birds was compelled to turn to other small food sources, which simultaneously were suffering from the increased use of pesticides and chemical fertilisers, causing widespread poisoning and reduction in numbers. The larger birds, foxes and badgers were forced to turn to other species for food — particularly game birds.

Brown Hare *(Lepus europaeus)*

Of the three species of hare native to the United Kingdom the brown hare is the largest. Unlike its relative the rabbit, the hare shuns the company of its own kind, tending to be solitary, preferring a habitat of lightly wooded and open country. A particularly handsome animal with an average weight of about 8lb, it has a russet-brown colour on the shoulders, neck and flanks, with its back often having colour variations of brown and grey, while its belly is snowy white.

Vegetarians, hares seem to enjoy a wide range of foods — carrots, turnips, cabbage, lettuce, young trees, flowers and, very occasionally, they will eat meat, invariably field-mice. Territorial animals, they stick to their own familiar area, though having the capability to range several miles. It is unusual for them to stray outside a fairly restricted area of several fields. They are good swimmers and seem to have no aversion to water, entering into rivers quite freely to gain access to a food source on the other side.

Most people have heard of the phrase 'mad as a March hare'. This refers to the often hilarious performances of buck hares during the breeding season, in the springtime. With apparently total disregard for their normal carefully singular lifestyle it is not unusual to see several bucks running and jumping over each other, boxing and kicking, often causing great tufts of hair to fly. During the period when they are engrossed in this activity they can often be approached quite closely, as they seem almost oblivious to anything but their own presence.

29

Sadly the brown hare is an animal which shooting men tend not to appreciate as much as I believe they should, all too often being shot as a matter of course during a shoot. This is unfortunate since, apart from a few localities, the brown hare is not particularly numerous. I do not want to give the impression that they are scarce, but rather to preach greater appreciation of these attractive animals.

Before shooting hares the sensible shooter should have an accurate idea of the number his ground holds, with his main aim being to use them for the table. They are not difficult to shoot and do not usually present much sport.

The majority of hares shot are flushed in front of the gun, springing from their sets and going straight away from the gun. This results in many being shot behind. As with geese, their large size tends to give the inexperienced the illusion that they are travelling slower and are closer to the gun than they actually are, so when shooting them swing well forward.

The hare's large eyes set on either side of its head give it an incredibly wide field of vision. They also have the capacity to act independently so that if a hare is being chased he can look backwards with one eye. Some years ago a friend and myself were walking 100 yards apart across a field when we each put up a hare. These two hares ran off down the same narrow track toward each other, both obviously looking over their shoulders at their individual sources of danger. Going full tilt they ran straight into each other, both bouncing high in the air, partially stunned. They gathered themselves together and whizzed off!

Hares have the capacity to breed throughout the year, but for obvious practical climatic reasons their main breeding period is through spring and summer, with several litters of leverets being produced during the year. Litters normally number four.

Blue or Scottish Hare *(Lepus timidus scoticus)*

The blue hare is smaller than the brown and is indigenous to the Scottish Highlands, blue referring to its colour change within the year. In the summer its coat is a brownish blue/grey, giving it an excellent camouflage on the heather mountains of its habitat. In winter its coat changes to white with the exception of the tips of its ears, which remain black. The formation of air bubbles in the fur gives the animal insulation to survive the fiendishly cold weather in which it lives. Needless to say when they are white they are perfectly camouflaged in snow, but this often leads to an odd scene when the animals have turned white before the snow has arrived. Thinking they are hidden, hares will often sit tight in their set allowing you to walk very close to them indeed. In areas where blue hares are particularly numerous these little white dots can be seen from a distance peppering the hillsides.

Because of the terrain in which it lives it is an animal which has for many years not attracted the attention of shooters, its main enemies being foxes, wildcats, and eagles, and it was commonplace up until the 1970s for most Highland estates to have a hare drive once a year, when willing, fit guns would be invited to the cull. Unsavoury affairs, they normally took the form of a line of walking guns driving large areas toward a line of standing guns. I have taken part in some of these shoots and it could be quite awesome as waves of these animals came pouring toward you, erupting from every clump of heather, and could result in huge bags of hundreds being shot in a day. This method of culling the numbers served to keep the population at a working level while giving the estate an annual cash crop from the game dealer. However, with the upsurge of interest in shooting in Scotland by foreign sportsmen most estates quickly saw yet another means of sporting revenue, since Continentals are prepared to pay fairly large sums to shoot these animals, and sadly the hare population of the high mountains has suffered as a result. Also, this new onslaught on their numbers, coupled with a particularly nasty series of winters experienced in the Highlands of Scotland in the late 1970s and early 1980s, served to reduce the population even more.

Blue hares have been introduced in several parts of the country, in Northern England, Ireland, and Wales, where they appear to do well if given peace.

Irish Hare *(Lepus timidus hibernicus)*

The third species of hare found in the British Isles is the Irish hare, which is found throughout Ireland, and is particularly common in mountainous areas. Generally larger than the blue hare the colour is more similar to the brown, having a more foxy or russet colour, and, although the coat does whiten in winter it is not as uniform as its Scottish cousin. A peculiarity of the Irish hare is that in the breeding season they will on occasion forsake their typical solitary behaviour and seek out others. They have been seen in great packs of about two hundred or more.

31

Grey Squirrel *(Sciurus carolinensis)*

The grey squirrel is regarded as being among the best-tasting small mammals on the American game list, and is very popular with pot hunters in that country. Strangely it has never been regarded as such in this country and no doubt the majority of British field sportsmen would never consider eating one. They cannot be considered sporting animals, but since any responsible shoot management would do its best to eradicate this pest it seems folly not to put its tasty meat to good use.

A native of the eastern states of North America, the grey squirrel was introduced during the nineteenth century as an appealing curio, no one at that time realising how it would eventually spread from Bedfordshire out in ever increasing circles until it is to be found in most counties in Britain today. Similar in habits and habitat to our own beautiful red squirrel, the grey is more aggressive and, although the two species' ranges overlap, the greys prefer to stick to deciduous woodland, avoiding conifers.

Although the grey is on the whole a vegetarian, it will eat a wide variety of foods — nuts, fungi, tree shoots, and small birds' eggs. It can be quite destructive in young woodland, nipping the buds from young trees and causing damage often quite out of proportion to its small size. With no natural enemies, once a pair enter a wood a whole colony can develop in a few years, and the red squirrels which may inhabit the same area tend to move on.

Grey squirrels have often been accused of being responsible for the reduction in the number of reds, but this is not entirely accurate or fair. The red squirrel population has been falling for a variety of reasons — the clearing of coniferous trees, disturbance, and the use of modern pesticides.

Feathered Game for the Shotgun

Pheasant *(Phasianus colchicus)*

If any bird typifies the British shooting scene it is the pheasant, found throughout the country, with the exception of mountainous areas. This bird is the mainstay of all large, low-ground shooting estates, and the aim of every small shoot organiser is to have a good head of pheasant on the ground.

Pheasants have been found in Britain since the eleventh century, and were possibly brought here by the Romans, but like so many

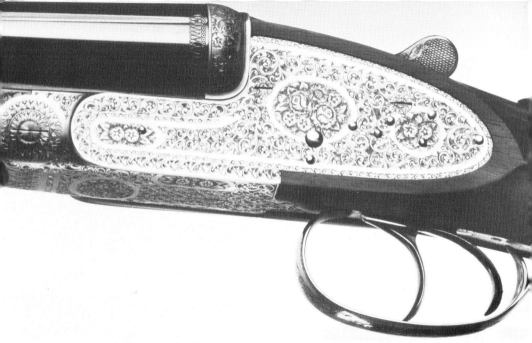

A fine example of beautiful engraving on a sidelock

A middle-of-the-range shotgun with attractive engraving

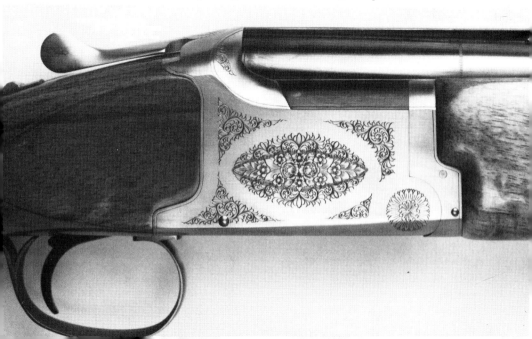

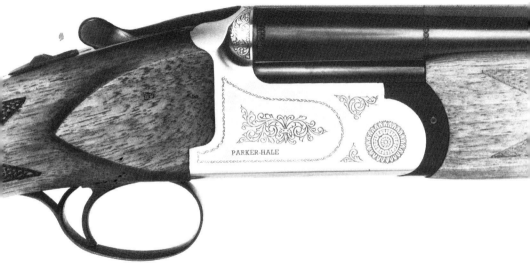

Plainer engraving on a less expensive gun

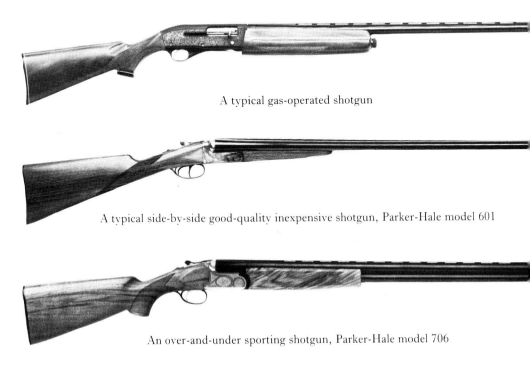

A typical gas-operated shotgun

A typical side-by-side good-quality inexpensive shotgun, Parker-Hale model 601

An over-and-under sporting shotgun, Parker-Hale model 706

other introduced species, their exact origins are unknown.

The many varieties originate from many parts of the Asian continent. Two varieties are to be found in the wild in Britain — the oldest pheasant known in the country, called the old English pheasant, a dark coloured bird with no neck ring, which originates from western Asia, and the variety known as the ring-neck, relating to the ring of white feathers around its neck. This latter variety began appearing in about the eighteenth century and originates from the eastern part of Asia. Both varieties are generally found in similar numbers.

Modern production of pheasants for shooting has been developed to a highly efficient degree, with vast numbers being released annually by estates throughout the country to cater for formal shooting. However, in recent years the economics of putting these birds in the air has taken its toll on the number of people able to afford this type of sport, and I believe that economic necessity will continue to erode the mass release of pheasants, forcing countrymen to have a greater reliance on wild stock.

The pheasant is an excellent sporting bird if driven or shot under the right circumstances, and equally can present a very unchallenging and poor target if shot in wet weather out of root fields, since wet plumage makes their flying less sprightly.

A particularly beautiful bird, the cock is about 30in long including the tail, with an iridescent copper breast, bottle-green head, large bright red face patches and a dark green rump before his long, barred 18in tail. Many colour variations can be found within one area. The hen has a shorter tail and is much less brightly coloured than the male. She can be anything from a dark chocolate mottle to light tan in colour.

The pheasant's nest is a scrape in the ground and poorly lined with local vegetation. The eggs, which can be up to sixteen in number, have a wide colour variation, from a very pale, milky-olive through to dark olive even in the same nest. The hen bird alone incubates the eggs and tends the chicks.

Their food in the wild is extremely diverse, ranging from plant seeds, leaves and fruit to most insects, caterpillars, and worms.

Anyone fortunate enough to have pheasants on his ground should carefully implement the control of vermin and do everything to encourage the natural population, as pheasants are a great attribute on any shooting ground.

Woodcock *(Scolopax rusticola)*

For many years there has been controversy which periodically rears up in the British shooting press over whether or not woodcock can carry their young between their feet or not. The answer is that they certainly can. I have seen this phenomenon with my own eyes on two separate occasions. However, I can also give a very credible explanation for the occurrence. When such incidents have been seen the birds have been sitting tight with chicks under, and when disturbed have zoomed off clapping up the undercarriage so quickly that a chick may become caught between the parent's legs. This may or may not be done on purpose, to remove the chicks from danger.

A beautiful bird about 13in long, the woodcock has russet plumage, a barred head, and long bill. The general appearance is rather squat when sitting. It is interesting to note that the woodcock's pin feathers were used as brushes by the painters of miniatures in years past.

Their preferred habitat is in wooded ground with plenty of damp areas, where they can search for earthworms, beetles, insects and grubs, which are their favourite food. Most woodcock are commonly seen during the 'roding' when the males fly at dawn or dusk in the spring or summer. They fly over woodland areas at tree-top height, in fairly leisurely fashion, giving a whistle call, which they sometimes alternate with a croaking sound. What they are doing is trying to attract the attention of hen birds. The hens are normally larger than the cocks, and the woodcock falls into the category of birds which should be shot sparingly, since they are not all that common and are pleasurable to see on the ground.

The nest is normally a shallow scrape in the ground lined with dead leaves. The four eggs have a base colour ranging from white to brown, and are heavily blotched with chocolate and grey. Only the female incubates, though the chicks are fed by both parents.

Grey Partridge *(Perdix perdix)*

The prettiest of the lowland game birds, grey partridges are widespread throughout the whole country, and can be found in most farmland and arable areas. They have suffered from a widespread reduction in numbers in recent years due to the increased use of pesticides and a series of wet springs, which cause a high mortality among chicks.

Today's British grey partridge is almost certainly the progeny of the mixing of our own native birds with the numerous imported birds that have been brought into the country over a number of years, from many European countries, particularly central European countries such as Poland and Hungary.

The grey partridge has a distinctive plumage, with grey underparts and a dark chestnut horseshoe mark on its breast. The base-brown wings are barred dark chestnut. It has a short tail and pretty head, both chestnut in colour, and large dark eyes. Hen birds are less distinctively barred than the cocks. Their food is predominantly grain and seeds, but is supplemented by insects, snails and spiders.

An interesting trait is the partridge's habit of roosting on the ground in groups known as coveys. They sit together in a circle, each bird facing outwards so that not only can they watch for predators but also in the event of disturbance each bird can easily explode away from danger.

The nest is a scrape in the ground, lined with grass and leaves, normally hidden in thick vegetation, containing anything from ten to sixteen eggs of a pale olive colour. Only the hen incubates, though the chicks are tended by both parents and can run and leave the nest a few hours after hatching.

Where partridges are bred from wild stock only, they should be shot with care, their overall numbers in your particular area dictating the bag. On the other hand, if they are released birds they may be shot accordingly. Partridges are good for eating, though their small size normally means that one partridge per person must be served.

Red-legged Partridge *(Alectoris rufa)*

The red-leg or French partridge was first introduced to Britain from France almost two centuries ago when, because of over-shooting, the native bird had been drastically reduced. The red-leg has become established throughout most of the southern counties of England where it is more numerous than the indigenous bird. Red-legs enjoy a habitat similar to that of the common partridge, though they prefer a slightly drier environment.

The two varieties of partridge cannot easily be confused. The red-leg has most distinctive black and white eye-stripes and heavily barred flanks. It is slightly larger than the grey partridge and does not have the horseshoe mark on the breast. The hen and cock have similar markings.

The nest is a scrape in the ground lined with dry grass and hidden in thick vegetation. The hen will lay up to about fifteen eggs, and often a second clutch of fifteen which will be incubated in an entirely separate nest by the cock bird. The eggs have a base colour of yellowish brown, with brown and grey flecks. The chicks are tended by both parents and they can run shortly after hatching.

Guidelines for shooting red-legged partridges are the same as for the common partridge. They too are excellent for the table.

Woodpigeon *(Columba palumbus)*

Whole books have been written about the shooting of this particularly sporting and tasty nuisance. It seems that the very presence of one or two landing in a field is sufficient to send the average farmer into fits of anguish, and woodpigeon shooting clubs have sprung up all over Britain, particularly in areas where alternative shooting is scarce.

This is the largest of the pigeon family to be found in Britain and is a handsome bird with its grey head, neck, and tail, the latter with a black tip. Most distinctive are the white patches on either side of the adult bird's neck, and when flying the white flashes on both wings are easily visible. The breast is a soft grey with a purple tinge. The sexes are identical.

Because of the pigeon's eating habits it has made itself the number one enemy of the farmer. In the early part of the year it eats clover, new-sown grain, peas, and other cereal crops, often congregating in enormous numbers, and its chicks or squabs are mostly hatched during August or September, with perfect timing to take full benefit of ripening grain fields.

The pigeon's nest is a simple platform of twigs, usually built in a tree, or a bush, and sometimes even on the ledge of a building or on the ground. It is usually fairly obvious, and contains two white eggs. Incubation is by both sexes and the chicks are fed by both parents. The hen usually lays two clutches.

The woodpigeon is one of the cheapest and most sporting birds available; but a little preparation is advisable for anyone intending to go pigeon shooting, in spying beforehand where they are currently feeding. A hide and a few carefully set decoys can bring great sport, the birds presenting virtually every conceivable angle of approach. The sharp, beady eye of the pigeon is an extremely good test of your ability to build an unobtrusive hide.

Pigeons are certainly the easiest birds to pluck, their feathers virtually falling out, and the meat is quite excellent. Since the greatest proportion of their body weight is on their ample plump breasts it is common practice, particularly if you have a number of them, to skin the breast, removing the two breast steaks only, discarding the little flesh on legs and wings, and of course doing away with the inconvenience of bones in the casserole.

There is no close season for pigeons.

Snipe *(Gallinago gallinago)*

Most shooting men's view of a snipe is normally as it zooms off in its typical zig-zag flight, or flashing past their noses as they wait at a pond for evening flight. The snipe is a pretty little bird with a brown-streaked body and striped head. It makes an unusual humming sound, especially during the courtship period of mid- to late spring. This is created not by its voice, but by the vibrations of its two outside tail feathers as the male sweeps through the air.

The snipe is to be found through the whole of the country, and during the autumn and winter migrant birds come to the country from the rest of Europe, which accounts for the fact that snipe seem to multiply during the winter months. It is the smallest of our game birds, about 10in long, the long bill making up approximately 25 per cent of the bird's total length. The bill is beautifully developed with a flexible tip for probing in mud and soft ground for the diet of worms, water beetles, grubs, and the seeds of water plants.

The nest is a shallow hollow lined with vegetation, and normally hidden in rushes, or long grass close to water. The little, pear-shaped eggs, normally four in number, are a delicate dark olive, boldly splashed with darker markings. They are incubated by the female only but both parents tend the chicks.

I believe the snipe should be shot sparingly. Although with their zig-zag flight they are extremely sporting and it takes a good shot to shoot one, I have always been of the opinion that being so small it is pointless to shoot one or two for the table, and unless your local population can justify the shooting of several, it is as well to leave them be and be content with having the pleasure of seeing them flying about.

Red Grouse *(Lagopus lagopus)*

Red grouse are native to Britain and occur only on high moorland heather areas in Yorkshire and some parts of Wales and Ireland outside their stronghold in the Highlands of Scotland. Their food is almost entirely heather, particularly young heather, though they eat different parts of the plant throughout the year.

They have gained popularity as the ultimate sporting bird because of their extremely fast flight when driven by beaters over carefully situated guns, and can provide quite electrifying sport, though sadly when driven they have become the exclusive game bird of the extremely wealthy. At the time of writing a week's shooting on a good grouse moor in Scotland for a party of eight guns can easily cost the staggering sum of £20,000. The red grouse cannot be bred in captivity, which adds considerably to its exclusivity.

Very territorial birds, the cock lays out a territory which he marks with a quite amazing display, leaping into the air, and running around with neck extended and tail fanned, challenging rivals. They are particularly susceptible, unless on well established, heathered ground, to parasites and disease, and require good quality keepering and husbandry if their numbers are to be kept up. The cock, 15in long, is a dark, reddish colour, with a most distinctive bright red eyebrow patch. Hen birds are slightly smaller and have darker brown bodies, which are more heavily barred.

The nest is a scrape in the ground lined with grass or heather and can have up to ten eggs, which are creamy white and covered with dark chocolate blotches. The eggs, like the grouse itself, are extremely hardy and seldom fail to hatch if the parents guard the nest diligently to foil hooded crows or other predators. Only the female incubates the eggs, though the chicks are looked after by both birds for about six weeks after hatching.

Ptarmigan *(Lagopus mutus)*
Ptarmigan occur in Britain only in the very high mountains in the Highlands of Scotland, where their habitat is restricted to the barren mountain tops. There they have to work hard to find their preferred foods of shoots, leaves, heather and the small fruits of mountain plants. They also eat a small number of insects.

The most common image of these birds is of the hen feigning injury to lure away danger, either human or otherwise, by dropping one wing, cheeping in alarm and running through the heather, trailing its wing as though it were broken. Similar to the red grouse, the ptarmigan is a territorial bird, but unlike the grouse they completely change their plumage colour. During the summer they have a tawny brown/red body with white wings, but turn completely snow white in winter, apart from their black tails.

Their nest is a scrape in the ground near rocks, and can contain up to ten eggs which are dirty white with dark brown mottles. They are incubated by the hen bird only, though both birds look after the chicks.

Ptarmigan can only be shot on the Highland estates under strict supervision since most estates have become conscious of their rarity value, and there is a responsibility not to overshoot them. Mostly they are shot only during high grouse drives.

Black Grouse *(Lyrurus tetrix)*

Fairly widespread wherever coniferous trees are to be found in the Highlands of Scotland, black grouse are also to be found in northern England and parts of Wales. The males are known as blackcock, and are at their most spectacular during the lek, which takes place in March, when they gather in groups to perform a most amazing display. The males, with their beautiful, lyre-shaped tails spread wide, jump up and down while running around displaying and fighting mock battles with each other.

The male is about 20in long with deep blue/black plumage, a flash of white on the wing, and white under-tail feathers. During the breeding season it has vivid red eyebrows. The females, known as greyhens, are considerably smaller, 14in long, chestnut and tan in colour, with a distinctive forked tail.

The nest is a scrape in the ground with up to ten eggs of a buff base colour lightly flecked red/brown. The hen alone incubates the eggs and tends the chicks. Their food is principally birch buds, conifer shoots, and a few insects.

There can be few more difficult and sporting birds than blackgame, shot under the proper circumstances — driven from a pine wood with the guns carefully positioned in a stand. Most estates with blackgame let the shooting, but choose carefully where you go since there is a tendency for some estates to fall into the temptation of taking in more guns that there are birds to be shot. The flesh of blackgame is similar to red grouse in flavour.

Capercaillie *(Tetrao urogallus)*

It would appear that the capercaillie had become extinct in this country by the end of the eighteenth century, and that the present population of birds has evolved from introductions in the 1830s to an estate in Perthshire. The capercaillie is by far the largest game bird to be found in Britain. They frequent coniferous woodland over a fairly wide range throughout the Scottish Highlands, with smaller numbers to be found in several areas in the lowlands of Scotland. They have not yet managed to establish themselves anywhere else in the country.

Capercaillie can be quite an awe-inspiring sight, their huge bodies weighing up to about 17lb and looking very similar to a turkey. The cock bird is really quite magnificent, with a great fan tail and bright red eyebrow, about 33in long with a general plumage that is grey-black with a brownish tinge to the wings, and a metallic, bottle-green breast. Hen capers are smaller, about 25in long, with a mottled red-brown plumage. The capercaillie's diet consists mainly of conifer shoots and buds (which makes them unpopular with foresters), though they also eat berries and some insects.

The nest is a shallow scrape in the ground lined with vegetation and containing up to eight eggs which are pale yellow flecked with brown. The hen alone incubates the eggs and looks after the chicks, the cock taking no part.

If you are tempted to try and shoot caper you will almost certainly have to go to an estate which advertises caper shooting. However, a word of caution is relevant. It is a sad fact that due to the enormous increase in interest in shooting in Scotland by wealthy Continentals, some estates have been tempted into taking in more shooting parties than the game could possibly support. It may seem immoral, and I am in no way saying that all estates subscribe to this policy, but it is true to say that of the large number of people who pay a great deal of money to shoot capercaillie only a small number are successful. In any case, unless you are one of that peculiar breed of trophy hunters, it is exceedingly difficult to justify shooting these magnificent birds as their flesh is not particularly palatable, probably because of the pine shoots which form the capercaillie's principal diet.

Tufted duck

3 WILDFOWL

Duck

There is a good range of duck which may be shot legally in this country during the shooting season, but the fact that a bird may be legally taken is not sufficient reason for it to be shot, and I advocate that sportsmen should restrict themselves to species which are more numerous nationally, such as mallard. It is really the height of irres-ponsibility to hear an individual congratulating himself on having shot a species which is generally unknown in his area. After all, it is every field sportsman's responsibility and duty to do his utmost to ensure that any species is given the opportunity to establish itself, and ideally the field sportsman should only ever 'cream off the top' of a species, taking sufficient for his domestic culinary needs and no more.

Wildfowl may be found in two main habitats — coastal waters and inland flight ponds or lochs. The wildfowler shooting on coastal waters should follow exactly the same basic rules laid out later in this book in the section relating to geese. Alternatively, if you intend to shoot them on a marsh, stream or pond where you know ducks to be flighting, get to your chosen area either at dawn or dusk in plenty of time to set up your hide.

All that is required for waterfowl is a small camouflage net screen interwoven with vegetation to break your outline and make you almost invisible to passing birds. If you intend to use full-bodied floating duck decoys on your pond, string them together in pairs or fours, 2-3ft between each duck. For pond shooting (see diagram) the string from the leading birds should be attached to the bank directly upwind, so that the decoys float clear of the bank, and bob about into the wind. For stream shooting, this is not always possible and the best method when using decoys in flowing water (see diagram) is to anchor the leading birds on a length of weighted cord, out from the

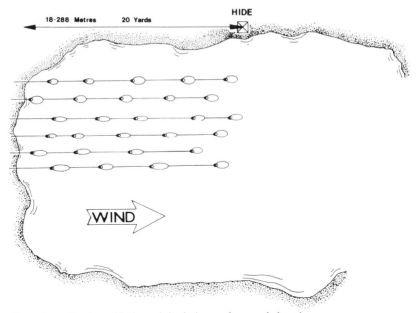

18·288 Metres 20 Yards

HIDE

WIND

Correct positioning of hide and duck decoys for pond shooting

bank, since if they are not in open water their silhouettes will not be seen by passing birds.

Most important when shooting over inland water is that you feed a pond you intend to shoot, establishing points where you sprinkle barley liberally around the water's edge and shallows, which will encourage birds to flight into it at feeding times. Then as long as you do not shoot the pond more than once a week, always take care to leave while the flight is still in progress, giving birds peace to get in at the food, you should happily be able to shoot the pond for the whole season. If on the other hand you are tempted to shoot the pond too regularly, or stay until late in the evening when the last birds have been shot at, you are asking for trouble in virtually guaranteeing that the birds will go elsewhere.

Mallard *(Anus platyrhynchos)*

Mallard are the most common of the wide range of duck found in this country. They will be found in virtually every wet area that is likely to hold a duck, from suburban ponds to Highland hill lochs. The drake is a handsome bird, with his distinctive plumage of green head, chestnut breast, blue/purple wing flashes and white neck ring. The duck, apart from blue/purple wing flashes, is brown, with colour

44

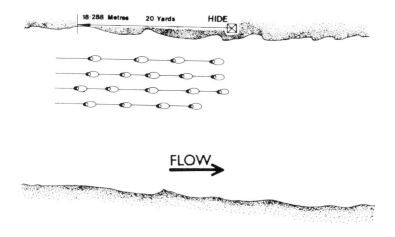

Correct positioning of hide and duck decoys for river shooting

variations ranging from milky chocolate to dark chestnut.

A most adaptable bird, usually it nests on the ground, though mallard nests are frequently found in trees. The nest is well constructed from local vegetation, such as leaves and grasses, with the cup beautifully lined with down and can be difficult to find. Up to sixteen eggs are laid in a clutch, with a colour range from a delicate grey to deep olive green. They are incubated by the duck only, and the chicks tended solely by the duck.

Their diet consists mainly of water plants and marginal seeds, though the adults prefer cereals such as barley and wheat, and during the autumn and winter will fly considerable distances to feed on stubble. The loudest and most typical 'quack, quack, quack' of ducks on a pond during evening flight is invariably mallard.

Other ducks that may be legally shot are teal, shoveler, pintail, common pochard, tufted duck, gadwall, wigeon and goldeneye.

Geese

There are many misconceptions about geese — their distinctive V-formation when flying for instance. It is not the case that skeins are led by old ganders. The reason they fly in a V-formation over longish journeys is that each bird gets a lift from the slipstream of the bird in front, and as the leading bird fatigues with the task of being out in front, it will fall back and another will take its place. Over a prolonged flight several birds can have the role of front runner. The

45

goose is the true master of communal long distance flight, and is careful to pick the correct flying conditions before migration, varying the height according to wind strength and direction. This results in the birds going particularly high with a tail wind, or dropping low, especially over water where they can get below the full strength of a head wind.

Pink-footed *(Anser brachyrhynchus)*

Pinkfeet are not quite as widespread as greylags, though they do occur in quite spectacular numbers. Their migration to Britain starts in mid-September, mainly from Iceland where they nest, and their build-up is continuous through to mid-November. They always return to the same favoured areas of the country, mainly confined to the northern and eastern areas of England, with particularly heavy concentrations in Scotland where they range throughout the whole of the central and eastern belt.

Easily identified in the air by their distinctive 'wink-wink' or 'pink-pink' call, they roost in marshes, estuaries and sheltered lochs, flying inland at dawn to feed. Their diet consists mainly of grass, grain, potatoes and turnips. Seen at close quarters the pinkfoot has a darker head and neck, and is generally lighter in appearance than the greylag. The legs and beak are a distinctive pink, with a blackish patch at the top of the beak, near the nasal cavities. Pinkfeet are extremely difficult to sex.

Greylag *(Anser anser)*

Greylags start to arrive in Britain from their breeding grounds in Iceland during mid-September and their numbers continuously increase through to mid-November. The species has a greyish-brown back, a pale-grey to white chest, often barred with heavy black bars, an orange beak and orange legs and feet.

The greylag is not common in England, occurring only locally in the north-west, south of the Solway, with small numbers to be found in the Wash. Their true home is in Scotland where, in favoured areas, they can be extremely numerous, often becoming unpopular with farmers whose crops they can damage. This is not usually serious, being mostly restricted only to loss of grazing. But in wet weather, if feeding on winter wheat, their feet can paddle down the mud, trampling and pressing the young shoots and causing crop damage.

One particular feature of the greylag goose is that of 'wiffling', a quite spectacular method of virtually vertical descent, when these beautiful birds seem to tip themselves straight toward the ground in a spiral. There can be few more exciting sights in the bird world than that of perhaps 200 greys wiffling down.

The greylag is the only native goose to breed in Britain, though in very small numbers. This is mainly in the far north of Scotland and the Outer Hebrides, though continuous efforts are made to encourage them to breed in other parts of the country. They appear to be intelligent birds that mate for life, with family groups staying together within the main bulk of the large flock. It is difficult to tell the sexes apart.

Canada *(Branta canadensis)*

The first Canadas were introduced to this country some 250 years ago, and were kept as ornamental fowl up until Victorian times, when a few escapees started to colonise lakes and waterways from the Norfolk broads southwards, and have steadily increased in number since. Large birds with a hefty ponderous flight, they have been a relatively easy target for anyone wishing to shoot one and this may have slowed the species' increase in numbers and range.

The sexes are alike, the most obvious feature being the glossy, black head and neck, with white chin strap extending up either side of the face to behind the eye. The general body plumage is a grey, black/brown with black tail and white underparts, stretching from the belly under the tail.

The nest, usually on marginal land, can be well hidden for such a large bird and is made of grasses and local vegetation, and lined with down. Up to six grey-white eggs are incubated by the goose only. The gander plays little part in the goslings' upbringing.

The diet is mainly marginal plants and grasses. As grazing birds they will often fly to grass fields, thus making them unpopular with farmers if they occur in large numbers.

Canadas are shot in the same way as other geese, though they are not difficult to kill as they are not so aerobatic. Perhaps shooters should have as their first interest the greater establishment of these delightful birds.

Wildfowling

There can be no more emotive and little understood subject than that of wildfowling in general, and goose shooting in particular. It is strange that some otherwise intelligent and experienced sportsmen seem to hold the most fanciful ideas about geese.

There are two basic types of wildfowling — foreshore shooting and inland flighting. As far as foreshore shooting is concerned, in many ways it is the most demanding shooting in this country, calling for a particular type of man who enjoys the physical exertion and is prepared to put up with the great discomfort he is likely to encounter. However, apart from the basic rules of shot size, range, and clothing, it would be wrong for any writer to lay down hard and fast rules regarding foreshore shooting, assuming that the shooter takes the necessary specialist equipment — a good compass and a flare or light stick. The latter is a plastic tube about the size of a fat ballpen full of clear liquid, inside of which is a thin glass tube with a different chemical. When you need to use it you bend the plastic which breaks the glass, the liquids mix and you have a vivid green light which will last for several hours and can be seen for miles. These are available in this country, mainly from firms supplying diving companies. There are also small pocket flares available which are equally easy to carry and anyone who embarks on a wildfowling expedition to tidal waters without one is being somewhat irresponsible, since every season we read of wildfowlers getting into difficulty.

Foreshore Shooting

Geese roosting out at sea fly at first light to their feeding grounds inland, and two areas may be picked — depending on the locality — to set up your position. The grass dykes and banks above the tidal water are obviously cleaner and safer than the tidal grounds, and if you have picked your flight path correctly then all you need do is dig yourself a depression where both you and your dog can blend into the countryside. With a small camouflage net it can be simplicity itself to set an ambush. Alternatively, if you are going far out on tidal muds you must be fully aware before you go of tide times and the speed of the incoming water. If there are no natural features which you can utilise — rocks, hollows etc — at your selected position then you must do your best to hollow out for yourself with a small spade, a depression in which you can hide, putting the sand in front of you like a wall. The real secret of any attempt to hide from game is for colours to blend and no movement to be seen.

As birds lift off and come toward you you must discipline yourself to take them well within range. Remember that on flat, featureless sandbanks it can be extremely difficult to judge distance accurately. If you are shooting at or near the water's edge a stone placed right on the tideline will indicate when the tide has turned according to how much is showing.

Inland Shooting

Scotland is the Mecca for serious shooters. Although this is the case, the situation for English shooters has changed dramatically over the last few years. Not only are the majority of landowners now aware of the potential revenue from geese, but also, sadly, a greater problem has arisen, that of the goose shooters themselves.

It used to be the case that shooting was relatively easily obtained in Scotland. This is no longer true, since over the last few years with the increase in shooting English and foreign sportsmen have made themselves generally unpopular as they arrive annually in their ever-increasing numbers, and understandably the Scots resent this intrusion. Scots are proud of their beautiful country and abundant wildlife. However, many sportsmen do not seem to realise this. Not only do they arrive on the foreshore areas in large numbers — to the great annoyance of the locals, but drive around the countryside looking for geese grazing in fields, and then approach the local farms expectantly asking if they may shoot the birds. I spoke to one farmer who told me that over a one month period he had an average of four car-loads of Englishmen a day in his yard. So if you wish to go goose shooting and you choose Scotland I strongly suggest that you book with one of the estates, and pay for the privilege of exclusivity. It is after all not expensive for the sort of facilities that the best of them lay on, including vehicles, hides, decoys, specialist knowledge and, most important of all, the fact that you should see no other goose shooters for miles.

As mentioned, geese start to arrive from Iceland and other areas in mid-September, building up in number continuously to a peak about the end of November. This is sustained usually until well after the season has finished, the last birds leaving in May. For many years I have been particularly delighted because not only do I see geese arriving while I am stalking stags in late September and October —

their small family parties coming over the high mountains in the latter part of their long journey south — but also because I have on several occasions, particularly during misty weather, had geese appear on the same level as myself, or even below me. Then in May, when I am stalking roebucks in the early morning, I see the great parties of geese making their way north again. If you go shooting in the early part of the season, though you may not see quite as many geese, the weather is more likely to be pleasant, and there is generally more to see for anyone interested in the outdoors. As the weather progressively deteriorates the amount of game to be seen diminishes, though goose numbers generally increase.

One normally flights geese at dawn and dusk, and the length of time over which these flights take place is governed by the weather and conditions prevalent on the day. Flights can stretch from half an hour to three hours of continuous activity. A number of people, normally those who are either inexperienced or just plain greedy or stupid, will sit all day under decoys, when the birds are hungry and intent on food, and that is usually when the big bags are achieved. Not only is it unfair on the birds, but generally against the whole sporting code. The question of how many to shoot is a difficult one. Personally I recommend that no one can justify more than two birds per gun in a flight. Also, it is a good discipline if you regard any bird that you have struck as one off your total bag. This serves to give the individual a greater appreciation of the birds, and an awareness of his own shooting frailties. I was once shooting with a fellow, and I asked him how many birds he required. He worked out that by eating one per fortnight for the rest of the year he could justify twenty-four. I pointed at a skein of about two dozen coming towards us and asked him if he wanted to be responsible for wiping out the whole skein from the sky. From the look on his face I knew that he understood what I meant.

Virtually every serious goose shooter has nightmare tales to relate of that most distasteful species of vermin — the marsh cowboy, or the sub-species — the inland cowboy. These are the half-wits who blaze away at birds at ranges a mortar would be hard-pressed to reach. I firmly believe that if all serious shooting men, when they witness such activity, did their best to remonstrate with them and point out the folly of their ways, the situation would not be so serious. They should be reported to any warden or authority of the area, or if possible to the BASC (British Association for Shooting and Conser-

Pheasants in a rearing pen

Partridges in a rearing pen

A grey partridge with bit fitted through the beak

vation, formerly WAGBI) and if they threaten you or give you cause for complaint — certainly report them to the police. I believe that if the more responsible of us who value our sport do everything in our power to stamp this out, things would quickly improve.

Many people who go goose shooting seem to think it necessary to carry special guns for geese. This is a misconception, and it is unnecessary specially to purchase magnums, 10-bores, and the like, since most people have false ideas about the alleged great increase of both range and pattern attained with these guns. You are obviously going to shoot better with the gun you handle most often, and that is your game gun.

Although it is now illegal to have more than three cartridges in a gun for game, very few men who own automatic shotguns can even begin to justify their use. If you are incapable of killing a bird with your first two shots, it makes no sense to loose off a continual volley, since the bird is fast getting out of range, and apart from the noise and cost, the possibility of wounding is high. The justification that some automatic owners use is that it helps them to get another bird. This is wrong, as you should be content with a few birds, and not intent on filling a sack. I have only ever met one game shot who carried an automatic, and could justify it — I have seen him take five pigeons out of the air, one after the other, but the same man refused to put more than two cartridges in his gun when he went fowling.

The best shot size to use is without any doubt No 3 followed by 1 or 4. It is ridiculous to use anything heavier, since you lose pattern. The argument put up by those who favour BB is that all it needs is one or two pellets in the right place. They never seem to stop and ask themselves what happens when you get one or two in the wrong place. If the bird is in range once you have mastered the speed and swing, it should be comparatively easy to bring it down.

Other equipment you will need includes a hide, decoys, goose call, and your ordinary shooting clothes, with warm underwear, since you will not be moving about. Take several changes of socks and underwear, as it can be wet and cold. It is also advisable to take a light, waterproof shooting jacket and waterproof over-trousers, and, of course, a vacuum flask.

As far as calls are concerned, there are a lot on the market, only two of which you need pay any attention to — the Scotch goose call, or the Olt. The Scotch call has a wooden handle which you hold in one hand, with a long, black corrugated rubber tube which you pump with the other. Remove the tube and throw it away, put the handle in your mouth and blow it. Not only will you get excellent results after a little practice, but it leaves both hands free. To my mind the best call on the market is one manufactured by P. S. Olt. It is small, neat, fairly easy to use and, hanging from a cord around your neck, is quickly got at and difficult to lose.

When using a goose call the secret of success is to use it purely as an aid to attract a passing bird's attention to your decoy set-up. If after a couple of calls the birds are showing no interest, it is unnecessary to continue since geese are motivated by hunger and if they are not attracted to come down for a look, blowing will be pointless. Most people misuse calls, and one often hears them blasting away like the Trumpet Voluntary. Two or three calls are sufficient if the goose has

Correct positioning of hide and goose decoys

WIND

Wal

any attraction to where you are. If not you are wasting your time.

If you do not have a well-trained dog which is totally steady, do not take him with you, as it will have to endure long periods of inactivity. Also, if the geese are coming in and your dog gets excited you are unlikely to get a shot.

Before building a goose hide it is essential first to study where you are going to place it. Since there are probably few birds quite as difficult to fool as the beady-eyed goose, arrive before dawn so that you can finish building your hide, set up your decoys, and be settled in by first light. Build your hide as small as you can, to accommodate you and your dog, and wherever possible try to get a good background — either a hedge or wall. Having erected your rods and net, gather as much of the vegetation from around you as you can,

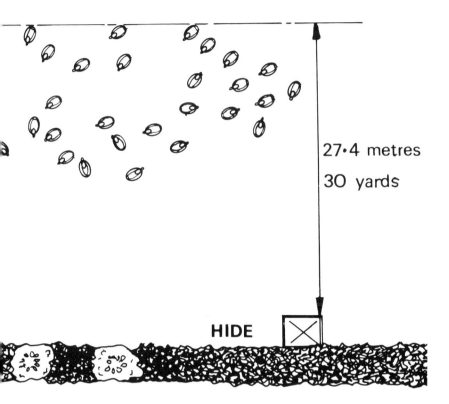

27·4 metres

30 yards

HIDE

dge

and thread it into the net. It is much better to pay that little bit more attention to the correct building of the hide than to discover when the birds start to come in that something is wrong or, worst of all, when you stand up to let your dog out the whole thing collapses! One must remember when you are in a hide and geese are coming in, that it is movement which they can detect, so you must freeze until the birds are in range, and keep your head down. If your face is above the level of the hide and you can see the geese, they can see you, and unless you are wearing a face veil, your face is flashing up at them like a great white beacon. It is best to keep your head below the level of the net and watch them through it.

I am often asked the correct way of laying out a goose decoy set up, and it is apparent that it is one area where the majority of shooting men do not think. Some goose decoys are particularly realistic, with lovely feathered texture finishes. That is purely to catch the buyer, and means not one whit to the goose. In America wildfowlers in some states lure snowgeese to land by pegging out sheets of newspaper, the white dots being sufficient to lure the birds down for a closer look. Most people know that milk bottles painted matt grey and laid in a pea field will bring pigeons whizzing down. So when trying to decoy geese the real secret is in the pattern which you create, as seen from above. Think about it. How ludicrous it would look if you laid them end to end to make a circle, or a perfect square. Something instinctive would tell the geese that all was not well. If on the other hand they are laid in an irregular pattern, as near as is possible to your own observations of feeding groups of birds, the pattern will look correct (see diagram).

Positioning decoys is of great importance. Geese will almost always land into the wind, though unless the wind is particularly strong, once they are down they will walk about, ignoring wind direction. When they are coming in to join other birds, nine times out of ten they will land behind the other birds, occasionally in the middle, and seldom in front. Therefore decoys should always be set upwind of your hide, bringing the birds past its front, with the last of your decoys almost in front of the hide (see diagram). The colour of the decoys is important, the white flashes of the tail being seen from some distance.

I have never found that a line of decoys on a flight pond makes the slightest difference to geese. If anything it may make them more wary as few individuals can successfully make the gobbling sounds

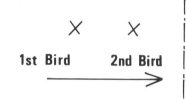

1st Bird   **2nd Bird**

Birds Flying in this Direction

The vertical line gives the position
where high birds, such as geese,
should be taken

that geese make on water. So if flighting a pond in the evening the main achievement of any goose decoy set-up will be to make you feel that you have set out your stall more attractively. However, the geese may not agree!

If you understand a little of aerodynamics it will help you to understand the importance of keeping still while waiting for the geese. As a goose is gliding downwards with its wings set, at the first sign of alarm all it has to do is tilt its primary wing feathers. This will instantly start lifting it fast in the air, without it bothering to flap. I have frequently seen, when watching birds dropping toward a hide, in the time it takes a shooter to lift his gun and swing, the birds climb an astonishing 30ft. To understand this tilting effect try driving a car at a reasonable speed with your hand out of the window. Keep your hand rigid, with your fingers pointing forward like a wing. At the slightest upward tilt of your wrist your arm will shoot up. To add a variation, if you tilt your hand sideways at the same time your arm will fly up at a sharp angle. Think how quickly a goose can move with two huge wings instead of a small hand.

'Swing right through the bird as though you were trying to miss in front', is the motto which should be adopted by all goose shooters, because the size of the goose makes it appear a great deal closer, and travelling much slower than it really is. Some of the most experienced game shots seem to lose all judgement when confronted by a group of geese either passing or dropping into a field. Very few of us ever miss in front, and if you hear the pellets hit the bird and it flies on, you are way behind. What you are hearing is the leading edge of your pattern striking the trailing edge of the bird's tail or wing

Duck decoys anchored on a flowing river

FLOW

feathers, and I think if you do this you should continuously remind yourself that the bird will almost certainly suffer from pellets in its body, which will at the least cause it great discomfort, and at worst kill it sometime in the future. This is totally against the sportsman's code.

Some writers and alleged experts who write in the shooting press do little to help, as I have read several descriptions of goose shooting in which the writers say that the struck bird wasn't hit hard; which is the same as saying you are a little pregnant! If a bird is hit at all, but does not come down, the shooter should endeavour to double his lead, because as I have already said, he is almost certainly missing behind. The other silly statement I have read in various sporting magazines is that the writer picked out the big gander. This is absolute rubbish since even the geese themselves can have difficulty deciding each other's gender.

All one need remember is that a goose is basically a large duck and if shot correctly will drop like a stone. Be content with a small number of birds; the day has gone when big bags were shot. After all the greater part of the pleasure of wildfowling should be in seeing the magnificent big birds, not killing them.

Flight Ponds

One of the most rewarding, interesting, and pleasurable attributes any shoot can have is that of a flight pond, not only increasing your sport, but providing a focal point and source of interest throughout the year. A flight pond can range from a simple wet area to a much

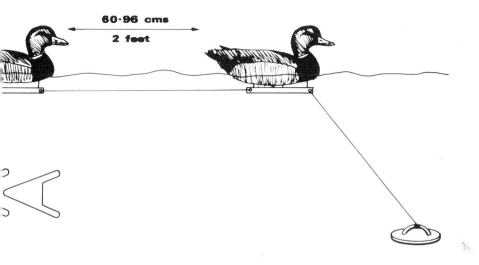

more elaborate affair containing fish, and if well planted and laid out can be an attractive long-term attribute to the area.

The amount of work that goes into a flight pond depends on a number of factors — the length of your lease, how much money you have, and how much land you can allow the water to take over. Whether you are creating one of the more elaborate affairs or a simple pond, the basic rules are the same, only the scale on which you are working varies. Although you can establish a watery hole, throw in food and get ducks, I strongly advise against that. Better to content yourself with the fact that a decent flight pond is going to take a bit longer. Ideally I would establish one in the autumn, letting it winter untouched to allow frost and wet winter weather to consolidate and pack the soil before planting in early spring. This means that your pond should be established for shooting the following season, and any problems that are likely to arise, such as water drying up, or increasing in volume, will have shown themselves to you. Using your imagination, try to create as natural an area as possible. A few trees, wired against rabbits, will add to its attraction, while giving more shelter. Stand back and plan the area. Do a few drawings, and try to think of the area and how it will look in years to come — is it large enough to justify an island in the centre (which greatly encourages nesting ducks)? Various aspects of your pond can be altered by the use of a shovel, but early planning is vital.

To make a flight pond is simplicity itself. If you are fortunate enough to have either a stream or boggy area on your shoot, most landowners will see the benefits of how such a feature can increase the amenity and value of the land over the long term. Your flight pond should not be too deep — no more than 20in is necessary. A couple of hours with a tractor and front bucket will soon scrape out enough top soil. Bank it around the sides and let your new pond fill with water. Whether you have diverted a stream or are relying on natural water seepage, let the water find its own level. Leave the site for a few weeks.

The best time to plant a few plants is either in the autumn or spring, and March is a good month. If you can get it, plant wild rice liberally around the edges of your pond for cover and food. Most waterways have an abundance of rushes, sedges, wild parsley and cress, and a few of these plants transplanted to your pond will give it a more natural appearance. A word with the owner of the waterway should easily gain permission to dig plants up for transplantation. Personally

I plant permanent hides, made up of young larch trees, in a semi-circle 30in apart. They quickly thicken up to make a perfect screen, though they will not be ready for the first two or three years. When they are thick all you do is keep them pruned to the size you wish.

By spring, when your pond is starting to establish itself, introduce a number of ducklings. Keep them in a vermin-proof pen for the first few weeks, then release them into the water. Make sure you give them plenty to eat to keep them in your pond. In fact the secret of keeping any game birds, whether it is pheasants or waterfowl, is to give them sufficient food, and most farmers are happy to supply barley at a reasonable cost. If he is shooting man or interested in wildlife, the farmer may even be perfectly prepared to share the cost of the grain.

Establish feeding points around your pond, sprinkling the barley both on the bank and at the water's edge. If you discover that this attracts other undesirable birds, then put the barley into the water at the edge of the pond, always putting it in the same place, and checking that it is all being eaten. Experience will teach you how much grain you need to put down. Ideally you want it all eaten since copious quantities left lying will sour the water.

Feed the pond regularly, keep the vermin in check, and you should have a shootable flight pond by the start of that year's shooting season.

Hide Building

Shooters, photographers, and game watchers must be able to blend into the countryside in as unobtrusive a manner as possible, without alerting the game to their presence. The best, most comfortable, and convenient method of achieving this is to construct a suitable hide. Hides fall into two categories — temporary ones erected for just a few hours before being dismantled and packed away, and fully permanent constructions to remain in place for long periods. All hides, however, should have common attributes — sufficient space for the occupant to move around, remaining as small as is practical, while still leaving enough space for such comforts as a seat, a corner for your dog, and room for you to lay out your equipment.

Temporary Hides
Practicality dictates how much equipment a sportsman can carry with him across fields. Shotgun, decoys, nets, camera, game bag,

and hide poles may not represent a particularly heavy weight, but are inevitably so awkward that after the first 100 yards you discover muscles beginning to ache where you never thought you had any. So therefore when setting up a temporary hide keep your equipment to a minimum and pack it with care. It is better to make two or three trips from car to shooting situation, but this is not always possible, because of the distance involved.

There are some good hides on the market and if you can afford it the simplest thing to do is to buy one of these ready-made ones. However making one up is easy. First of all you will need a piece of camouflage net. Avoid ex-military netting with hessian strips, as these become extremely heavy when wet, and if not thoroughly dried out when put away will rot and mildew fairly quickly. Better to go for the lighter netting made of nylon, with small nylon or plastic camouflage strips or leaf patterns attached. (Some modern military netting of the same construction is available.) These nets are light even when wet and can, if necessary, be stored away in this condition without damage, though I do not advise that they be left beyond the first dry day as they will start to smell. The length of net you require is largely a personal choice dependent on the amount of space you need, but should be a minimum of 8ft long by 5ft deep. Six-feet long sharpened broom shafts are excellent as uprights, since they are cheap and replaceable. Matt green-painted tent poles are also very good, though normally requiring guy-ropes. Best of all are made-up metal rods with a foot press (see diagram). These can be welded for you in any blacksmith's shop and will last for years. Four strong guy-ropes and a dozen metal tent pegs will complete your hide materials.

When building a temporary hide it is best to put it against a good backdrop — a wall, trees or bushes. Push two rods into the ground against your backdrop, 5ft apart, and the other two parallel, 4ft out into the field. Then starting at one back corner, take a couple of turns of your net around the rod, out to the next rod, giving it a couple of twists and so on, around all four rods. If you have placed your net against a thick thorn hedge, netting at the back should not be necessary.

With all four sides up, guy-out your rods and peg them down to make them rigid. Then gather any excess netting, pulling it out at a slight angle, and peg it in. This will prevent it flapping about in the wind. Always make sure that you leave one side specifically as the door so that you can pass in and out of the hide without dismantling

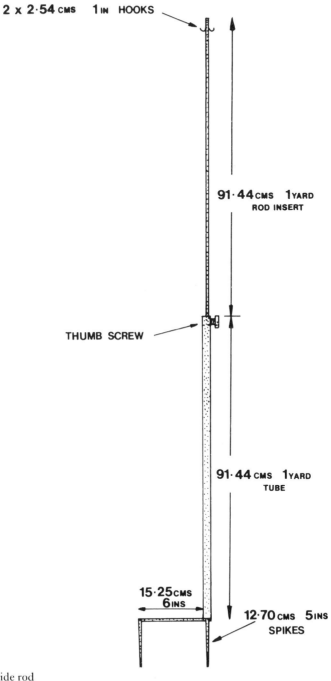

2 x 2·54 CMS   1 IN   HOOKS

91·44 CMS   1 YARD
ROD INSERT

THUMB SCREW

91·44 CMS   1 YARD
TUBE

15·25 CMS
6 INS

12·70 CMS   5 INS
SPIKES

A home-made hide rod

it. Establish one side for yourself — seat, equipment and so on, and the other for your dog. That little bit of extra care at this stage is wise, avoiding a situation, when the action starts, of you tripping over your seat or dog.

Gather up some handfuls of vegetation and thread them into your hide through the net. Experience will teach you how much to use since you neither want a haystack, which would be obvious, nor a hide that is too thin. If you stand back and look at your work you will see the spaces that should be broken. Remember that a hide which is completely solid is as noticeable as no hide at all, and the function of the camouflage net, apart from the fact that it blends with the background, is really to break your own outline and conceal movement.

Permanent Hides

Permanent hides can be situated on the ground or in a tree, and should not be confused with high seats. A permanent hide is normally used for photographic work, when, over some months, you want to keep a record of activities surrounding a bird's nest or animal den, and, depending on how long you intend it to be there, can be fairly substantial. A ground-based permanent hide is best built with fence posts and rails for the frame outline in the size you require, with stretched quality nylon camouflage netting over it and tied down. Inside a permanent hide you can build a solid bench to sit on and to lay your equipment upon. In other words, following the same rules for the construction of a temporary hide. You simply make it more secure. The best possible permanent hide that anyone could wish for is one created, if the vegetation is conveniently growing where you want it, out of growing bushes. By a bit of judicious and careful work you can, by pruning the inside of a gorse bush for instance, create a perfectly natural yet impenetrable screen which will change colour with the seasons and be completely undetectable.

Tree Hides

Whether you intend to shoot or photograph deer, if you do so from a tree there is no need to go to a great deal of trouble to make yourself invisible, but better to go for a secure structure with a solid platform and handrail in the branches, since deer generally do not look up, danger never coming from above. On the other hand, if you wish to photograph a nest, squirrels' drey etc, then obviously you must use the same tactics as ground hides, making them more difficult to detect. But whatever you do, the basic rules of tree hides and high seats are that there must be a safe method of getting up to them, and they should be secure, safe, and comfortable since you may have to spend long periods in them. I know a fellow who has three seats on his estate, and so determined is he for comfort that he has gone to the trouble of taking old armchairs up into the trees. This may be slightly unnecessary, but his comfort is definitely assured.

Stoat

4 VERMIN

Anyone who is seriously interested in good game management must control the numbers of species that prey upon game. However, let me make it absolutely clear: it is my firm belief that predatory birds — falcons and hawks, etc, — must not be included in your general game plan. The responsible field sportsmen of the 1980s must welcome these birds and accept that, though they do a little damage, to kill or disturb any of them is to my mind extremely short-sighted, selfish, and immoral, apart from the fact that it is also illegal. It is much better to value the beauty of the peregrine wheeling across the sky. Accept that he must eat a few grouse and that any small number of game birds lost is little price to pay for the privilege of knowing these birds are doing well.

A lot of absolute piffle has been believed for so long about our other predatory species, and far too many individuals of the old school still believe that sparrowhawks, kestrels and buzzards are harmful. They certainly are not, and must be valued by any true countryman. However, there are other species of birds which must be controlled, otherwise they are capable of doing considerable damage; birds such as black-backed gulls, hooded crows, and the crow family in general — with the exception of rooks which, in the main, do not do anything like the damage they are believed guilty of.

The real problem species for shoot managers are four-legged — cats, foxes, stoats, mink, rats and weasels (to a lesser extent), and it is pointless to do any work increasing stock, introducing birds, and making general improvements if the individual does not work equally hard in the control of vermin. The great success of all the four-legged predators is their keen sense of survival; they are usually aware of your approach and have disappeared before you know they are there. However, they cannot prevent themselves leaving, to the trained eye, distinct evidence of their presence, in their tracks, droppings, and often their half-eaten meals. Obviously snow is ideal to reveal tracks of every variety, but soft mud and soil around river banks and flight ponds, farm tracks, and other areas where the ground is soft may be checked for evidence of pests. The identification of areas where these animals live and move about is of first importance before any form of trap is set to capture them.

Forgetting for a moment the illegality or otherwise of killing certain species, most of us know that the chances of being caught in a remote country area killing a badger or otter are low, yet that is not the point. Far too much humbug and silly old wives' tales have been spread about such creatures as badgers and otters, and, as I have said previously, predatory birds. It is every responsible outdoorman's and woman's duty to share the environment with these relatively harmless animals, and enjoy their presence. Yet step very firmly on the pests already mentioned, for they are the groups that must be kept down.

But one must be careful not to upset the balance of wildlife, for by removing one species another might flourish. A prime example of this was the back garden of my mother's home in a Glasgow housing scheme which bordered the countryside. My mother, for a considerable time, delighted in the fact that a stoat was living in her back garden, its pretty, sinewy body often being seen nipping through the rockery from one garden to another. Then one day tragedy struck, an idiot youth living next door shot the animal with his shiny new air rifle, and the occasional rat began to appear. Shortly afterwards the neighbours were forced to call in the Department of the Environment to lay their poisons. The removal of one species had allowed the upsurge of another. It was obvious that the stoat had only taken up residence because of the ready food supply of rats and mice, and after its demise these other creatures flourished.

Foxes (Vulpes vulpes)

Before the advent of myxomatosis the fox's main diet was rabbit. However, now that rabbits are no longer as easily and widely found, foxes eat rats, mice, voles, hedgehogs, beetles, snails, and virtually anything edible, including grass, bulbs, fruit, and root vegetables. Ground-nesting birds are particularly vulnerable since they are easily caught. When these birds are nesting both the dog fox and vixen are feeding cubs (born in March and April) and foxes must be kept down. The best method of doing this, if you can find their earth, is to use trained and experienced terriers, or alternatively gas them with cymag gas if you can get it, carefully following the instructions for use. Fox drives can be successful, though the guns must stay close together, be experienced and safe. I do not like the setting of snares or the use of poisons when dealing with foxes, since both are indiscriminate and outside your own control. However, if you are quite certain that a hole in a fence is being used by foxes, and not badgers, then you can snare them by using one of the heavy wire fox snares.

Foxes prefer areas of cover, woodland and hedge bottoms, and only at night are likely to wander across open fields. Their tracks are very neat, with all four toes and claws normally visible, the fore feet being slightly larger than the hind feet and almost in a single line when walking or trotting. Depending on the size of the animal, the stride varies from 40 to 50cm, though when moving at speed this distance could extend to 60cm or more, with all four feet placed in a cluster.

Wherever a fox has established an earth in a hole in the ground, bank, under a tree, or in a pile of rocks, you would normally find chewed bones, tufts of fur and feathers scattered about and, if the fox has been a recent occupant, the pungent musk smell which is given off by the anal scent glands. Their droppings resemble a twisted piece of thick tarry string, with tapered or pointed ends, anything from 3 to 7cm long, and can be coloured from dark green to blue-black, depending on the food content. If examined they should show pieces of bone and compacted hair, and, if reasonably fresh, the droppings will give off a strong musk smell. Other signs of a fox's presence are small tufts of red hair on holes in fences, and scratch and digging marks where soil may have been disturbed either at the entrance to a burrow, or where the fox has dug up a mouse.

Domestic Cats *(Felix domesticus)*

Cats are a touchy subject and one of the areas where the field sportsman must use his intelligence and discretion, particularly if he lives in an urban area, where people's pets are likely to be caught. All cats are hunters, and capable of creating death and destruction far out of proportion to their size. A cat is much more destructive on your ground than a fox. Few cat owners would believe that their domestic tabby when out of sight in a thicket becomes one of the most efficient killing machines nature has evolved. Cats therefore must be

A French partridge

Mallard ducks in a release pen

Goose decoys: half shell *(left)* and full body *(right)*

shot whenever they appear on your shooting ground.

Even the most gentle domestic housepet will, as soon as it goes outdoors, display all the abilities that have made cats among the most sophisticated and deadly predators in the animal kingdom, and they can be extremely difficult to locate on a shoot. They will not, for instance, follow paths, but slip quietly across them. They have no regular runs, and prefer to move about concealed. However, their tracks are easily identified with the pad having three distinct connected imprints, with four toes in an arc in front. No claws are visible. The tracks are neatly placed on the median line when walking, and when the animal is stalking each track would normally touch the one in front. Part of the cat's great stealth is that each foot comes forward touching the one which is already firmly placed, leaving connecting prints. Obviously the size of prints depends entirely on the individual cat, and with such a wide size variation in domestic cats it would be inappropriate to give track dimensions. Although their droppings tend to be banana-shaped with pointed ends, it is not normal to find them since cats bury them as part of their desire to remain undetected.

Cats have a particular habit when eating an animal such as a rabbit. They roll the skin back as they eat, giving the impression of it having been turned inside out. A cat will not normally finish a large rabbit in one sitting, but having eaten its fill, will then cover the carcass over with leaves and grass to hide it, and return later. If there are no flies about, such as during the winter, individual cats have been known to return to their prey up to ten days later.

Stoats *(Mustela erminea)*

If a list of the least desirable beasts to have on shooting ground was drawn up, it would be topped by the stoat. These beautiful little creatures can do a great deal of damage to young game birds, nesting birds, and eggs of all bird species. If given an opportunity stoats will flourish, their numbers rising at surprising speed.

The stoat prefers walls, ditches, or hedge bottoms, anywhere in fact where it can find close cover where it can hunt about for its favourite foods which are diverse and include birds, small mammals, reptiles, rabbits, and even hares. In the case of animals such as rabbits or hares the stoat fastens itself to the back of the neck and either bites through the back of the skull or down into the vertebrae at the back of the neck.

One well developed trait of the stoat is that of curiosity, which can be their undoing if you wish to trap them. Confronted with any sort of little hole, burrow, or tunnel, the stoat will be compelled to explore its recesses. I was once sitting early in the morning, leaning against the base of a tree in an oak wood near my home, waiting for a friend. I spied a stoat working its way through a whole series of disused burrows. First it would go down one to reappear at another, its curious little face peering around. Then it would dart down another hole to reappear some minutes later at yet another exit. All the time the little creature was moving toward me, and as I sat motionless it popped out a couple of feet from my outstretched legs. It had a good look at me and then to my surprise hopped out of the hole and ran across the grass toward my boots, which it sniffed with great curiosity. It was only when it occurred to me that it might scuttle up my trouser leg that I made a movement, and in a flash it was gone.

The stoat's presence is easily revealed by the carcasses of what it has been eating. Most typical is that of a rabbit, with the back of its neck and the flesh of its shoulders eaten. Being small creatures their capacity for food is obviously limited, and they almost always leave plenty of signs. Similar to its American cousin the skunk, stoats produce a very pungent, foul-smelling musk from their anal scent glands when frightened. Their droppings are like small pieces of tarred black string which smell strongly of musk, especially when fresh.

Their tracks, normally only found in soft ground or snow, are similar to all those of the weasel family. The hind feet have four distinct little pads with the toes and claws showing around the top. In very soft ground webbing can be seen. The track of the fore feet is of three pads, with five little toes and claws set widely apart. The front print is usually about 2cm long and the rear about 4cm. When stoats move about quickly all four feet make grouped prints, which could be anything from 18 to 24cm apart, though when moving at top speed the groups of prints can be up to 50cm apart. In winter stoats change colour, turning snow-white except for the tip of the tail which remains black. In this condition they are referred to as ermine.

Weasels *(Mustela nivalis)*

As far as the game manager is concerned weasels are really miniature versions of the stoat, but are so small as not to prove any real threat to game. Their main foods are mice, voles, and all the small mammals. They will of course take birds' eggs, but would be hard-pressed to do any real damage to game birds, other than those similar in size to quail, and should not therefore feature too heavily among your persecution priorities.

Weasels are widely distributed throughout the whole of Britain, but because of their habitat, tiny size, and mainly nocturnal activity they are rarely seen.

With the same arched-back running movement common to the entire weasel family, their little tracks will be found from 20 to 25cm apart, under normal speed, and up to 30cm between each little group of tracks when moving at high speed. Their tracks are almost identical to those of the stoat, but very much smaller, the hind foot being 1.5cm long and 1.2cm wide. The fore feet are 1.2cm long and 1cm wide. The claws are almost always shown. The droppings are also similar to a stoat's, though the ends are more pointed.

Mink *(Mustala vison)*

By the mid-1960s mink had become established in every county in Britain, and their population has been increasing dramatically in some areas ever since. Mink will always be found not far from waterways since they are excellent swimmers, and their preferred habitat is normally beside rivers or lakes where they do most of their hunting. They will take a wide variety of fish, reptiles and all waterfowl. They can also be destructive to nesting birds, taking not only the eggs but also the sitting bird.

Their tracks are easy to distinguish, the fore feet made up of five toes, normally with claws clearly visible, set around a central irregular-shaped pad, while the hind feet show four toes set around the pad. At normal speed their tracks, in groups of four, can be widely irregular, though when moving at a consistent gate, the distance between groups can vary from 35cm to 50cm. The droppings are sausage-like with the same twisted, tarred-string appearance as that of all other members of the weasel family, and are 6 to 8cm long and approximately 8 to 10mm across. The droppings would usually contain evidence of the animal's food — particles of bone, hair, fish scales and so on. A particular feature of them is that they are foul smelling. A hole where a mink is currently in residence would have droppings near the entrance, and the unpleasant and distinctive musk smell would be present.

Rats *(Rattus norvegicus)*

The brown rat is a creature found almost everywhere in the British Isles, though it was introduced to the country only in the nineteenth century. There is a country saying that where you find men you will find rats. During winter in a country environment rats are drawn to human habitations where they can capitalise on the warmth and shelter of buildings and where there are easier pickings to find. As the weather warms up the rat begins to gravitate back to the country, making its home in any convenient spot — walls, rabbit burrows, riverbanks.

The rat has the capacity to adapt and eat virtually anything it can find — vegetable matter, seeds, insects, birds' eggs, and small chicks, and every effort should be made to eradicate them wherever they appear since they can multiply at alarming speed. Some years ago a friend of mine dumped piles of potatoes and barley in regular heaps along the riverbank on his ground, a considerable distance from any habitation. Some weeks later we were both walking along the bank and were surprised to see a large number of rats scuttling about in daylight wherever the foodstuffs had been dumped. They must have been drawn from a fairly wide area.

Destroying rats is easier around buildings than in open country. Within barns and outbuildings tracking powder has good results if put in their runs and entrances to their holes. Picked up on their feet it is transferred to their mouths as they groom themselves, with fatal results. Poison barley is particularly effective, though care must be taken when distributing it, making sure that it is placed in situations — such as the inside of walls — where no other foraging animal or bird will pick it up. If there are no walls or other convenient receptacles for poisons then a series of cage and fen traps must be used. However this method is much more laborious, and is more likely only to contain rather than eradicate the infestation.

Rat runs and burrows are found in the soft earth around riverbanks and their track is fairly distinctive. The fore feet leave a broad print, with four toes normally visible, between 1.5 and 2cm long, and 2 to 3cm wide. The hind feet have three middle toes in a line to the front of the track, with the first and fifth toes splayed out at a more acute angle, 2.5 to 4cm long, and 3cm wide. Occasionally marks reveal that the tail has been dragged, though this is not common.

Snaring and Trapping

The most important rule for anyone who intends setting traps of any kind, whether a snare or fen-type crushing trap is that it must be checked daily. Correctly set traps or snares can be a most effective way of killing vermin, though it must be stressed that it is illegal to set a snare or any other trap which is not checked at least once in every twenty-four hours. Personally I recommend once in every twelve. This means that you have more chance of getting two rabbits from a snare in a day, at the same time greatly reducing the risk of your quarry suffering unnecessarily. Never mind the fact that every time you find an occupant in your trap it is already dead. It is your absolute duty always to assume that the animal may still be alive and to do everything in your power to reduce any chance of it suffering to an absolute minimum. Incorrectly set snares are a waste of time and cause lingering pain.

Never set your snares in a field which holds stock. Inquisitive cattle and sheep have an unfortunate tendency to prod with their tongues at curious objects, particularly if they have a strange smell, and, caught by either tongue or lower jaw, hideous wounds can result. The real art of snaring was perfected in the days before wire fences, when rabbits were more abundant and keepers and countrymen would snare the runs in fields. Any keeper worth his salt (even today) would be extremely cautious about setting a fence snare since they invariably take as many pheasants and partridges as rabbits.

The best time to set snares is on a dark windy night. Rabbits, in common with many other animals, are more skittish in wind and run better. Also, there is less chance of them hearing the squeals of any occupant of other snares in the line. This is particularly true with foxes. On a dark, rainy night foxes are not quite as alert and are more likely to walk into a snare than they are on a clear, still, warm evening when they are enjoying the hunt.

When laying out a snare line it is best to mark every tenth snare, checking them by number as you move along the line. This means that if one is missing you know exactly which section to check, and remember you cannot snare a march or boundary fence — even for vermin — without the permission of the landlord on the other side.

Pin Snares

A pin snare for a rabbit on a run should be made pear-shaped. The bottom of the loop should be 4in from the ground, with the noose 3in across. The eye of the snare should be at the bottom, so that in the event of it being hit by a passing bunny it will spring back into place, with the peg firmly hammered into the ground. Do not fall into the temptation of 'that will do'; always err on the more secure side. In the long run it pays off.

Fence Snares
Once you have satisfied yourself that a hole in a fence is being used by rabbits or perhaps cats, and not game birds, the correct way to set the snare is to suspend it from the wire above, again pear-shaped, but this time with the eye at the top. The loop should be 3in across, with the bottom wire 4in from the ground, and the snare must be firmly fixed.

Fox Snares
The most common location for the snaring of foxes is normally at a hole in a rabbit fence where foxes are known to pass. However, you must make absolutely sure that the same hole is not being used by either badgers or roe deer. Most people imagine that deer always jump fences, yet on many occasions I have seen roe squirm through a small hole in a 3ft high rabbit fence they could have hopped over effortlessly. So if you are in any doubt do not set snares, remembering at all times that they are indiscriminate. Also, when picking a position to set a fox snare you must be careful as to your methods of securing. If you are using a pin it must be really robust, and hammered in flush with the ground, otherwise an animal which is not caught around the neck can chew the head from the pin and disappear with the wire still attached to its body. If the ground is too stony it is better to secure the snare to a heavy weight which an animal cannot drag. I have found fence posts to be ideal, particularly near plantations, since even in the event of a large animal being caught by accident, such as a deer or even a dog, if it manages to drag the post it would snag itself at the first brush or trees.

Fox snares should again be pear-shaped with the bottom of the loop 6in from the ground. Foxes travel with their heads held low, and the opening of the loop should be 7in across. Any larger and the animal may be able to get its shoulders through the loop, which must be avoided. Try to refrain from handling snares with your bare hands, since the less of your scent they have on them the more productive they are likely to be. The only snares which you may legally use are those which have the running eye or noose system. The self-locking snare is now illegal. The noose system has an unfortunate tendency to relax when the animal stops struggling, while still holding it, so you must attend the trap at least once a day and and check it. The best type of snare to use is one with a swivel, which prevents the animal twisting the wire until it breaks.

Tunnel Traps

Stoats, weasels, and rats have a preference for running along the bottom of fences or walls and these are the ideal places to lay your tunnel traps, which are easily constructed. A small fen trap is set and laid at the foot of the wall covered over with two lines of bricks, with slates laid over the top to create a small dark tunnel which these inquisitive animals will want to investigate. A better method than the brick and slate covering is for you to build a tunnel using a wooden frame made of two batons of wood approximately 20in long with a marine ply top, the tunnel made to give sufficient clearance for the jaws of the trap to snap closed. I always wire down the trap chain to prevent larger creatures, mink for instance, from dragging the trap if not caught around the torso. Also, it is a fact that accidents can happen, and it is much better if you have caught someone's cat, that you arrive to find it secured, than the cat dragging the trap homeward with the ensuing bad publicity that would almost inevitably develop. Also, an animal like a cat dragging a trap invariably gets it caught fast in vegetation, quite often some distance from where you expect it, and if not found the animal suffers a lingering death. It is therefore preferable to wire it down.

Mink Traps

If you have a stream or pond on your shoot it is a good idea to assume you have mink whether you have seen any sign of them or not, since these animals are spreading rapidly throughout the country, and travel on waterways. Setting a mink trap once a fortnight is sufficient precaution against an animal moving in on you. If on the other hand you have an established mink population already you must set the trap daily, or as regularly as you can. The problem with setting a trap fortnightly is that you must remember to check and when necessary empty it the next morning.

By far the best traps to use in the control of mink are the large cage traps sold specifically as mink traps, which have a neat trap door system that allows the animal to go in for your bait — a piece of fish or rabbit meat — snapping closed behind it and catching it without harm. This means if a young otter has blundered into it you can release it easily. On the other hand, if you have caught a mink in the trap it is best to shoot it while it is still caught in the trap. Do not attempt to take it out alive as mink can inflict a very vicious bite. Shoot it with an airgun at close range in the head.

Cage Traps
Cage traps, although expensive, are far preferable when attempting to trap animals in open and exposed areas such as riverbanks, and it can be quite amazing the number of rats that appear where the shoot owner would swear none exist. These traps are exactly the same as mink traps but smaller and thus cheaper.

Poisons
In recent years there has been an alarming increase in the use of poisons throughout the country with people using undiluted crop dressings injected into eggs and dropped onto flesh. Irrespective of how efficient you are told these chemicals are you should not use them. Not only is indiscriminate poisoning illegal, but the compounds are often so persistent that they are virtually impossible to destroy and remove from the food chain, killing both predator and prey.

Crow Traps
The most effective legal way of catching members of the crow family is that of a cage trap, and it has been my experience that to be truly effective their location should be changed periodically, since these birds appear quick to learn. So such devices in set locations should be avoided.

Brown hare

5 GUNDOG TRAINING

When considering buying a gundog the first thing to do is examine the type of shooting you do and decide whether you really need a dog. Owning one brings a great responsibility since with luck, guidance, and good food the animal should last for ten, twelve, or more years. Therefore a great deal of thought must go into it.

Having decided that your sport justifies a dog, the next step is to decide what sort of dog is right for the job. There is a wide variety of gundog breeds that can be purchased throughout the country; some common, such as the labrador retriever and the English springer spaniel, and other lesser known breeds, for example the Irish water spaniel and the munsterlander. The breed of dog you choose will depend largely on the type of shooting you do. Are you basically a rough shooter or a wildfowler, or do you enjoy the minority sport of moorland shooting? For whichever shooting you do there is a dog which will suit your needs and circumstances, and while most well-bred dogs from a good working strain are capable of working in all shooting disciplines there is little point in buying, for instance, a small English springer spaniel when your penchant is for shooting the large and heavy greylag goose. Far better to get a big, strong thick-coated labrador for that job.

The English springer spaniel is a hunting dog and remains the most widely used and popular rough shooting dog in Britain, for good reason. The springer is most at home bustling around in front of the gun, flushing or 'springing' game out of every bush and thicket, game which would lie undetected by other breeds of dog. Shooting over a good springer is both pleasurable and gratifying as it flushes out pheasants from all corners of a root field.

The original springer spaniels were used only for hunting, and were not asked to retrieve until the beginning of this century when the desirability of a more versatile all-round dog was called for. Springers are generally not as capable as the labrador at performing difficult retrieves. However, the springer is an efficient hunting and retrieving dog on both land and water and will not baulk at entering dense cover. Its thick, long-haired coat, while protective in cover, does attract dirt easily, and unless bathed regularly the springer can have a distinctive canine odour.

Wildfowling probably makes the greatest physical demands on both the shooter and his dog. The weather conditions are usually bad — cold, windy and damp, on top of which the wildfowler's dog is expected to work in cold water. The ideal choice of dog for the wildfowler, or anyone who shoots regularly over water, is the retriever, in particular the labrador. As the name suggests, the retriever is a master at retrieving and will quite happily go out on long distance retrieves if asked. Although not in the same league as springers at hunting, labradors can certainly perform this task adequately.

Labradors were originally developed (probably in Newfoundland and not Labrador as the name suggests) as working dogs for the fishermen of the north-eastern coast of America, pulling heavy ropes from boat to boat. The labrador was used also by the commercial wildfowlers from that part of America and often had to make as many retrieves in a day as some of us today would expect in a season. The true labrador has a thick double coat to keep it warm in arctic conditions and the distinctive short tail known as an otter tail. These dogs can cope with a long and arduous day's work in dreadful conditions. Unfortunately there is a strain of working labrador today — the trialing dog — which bears little resemblance to the labrador as most of us know it. It is a small, thin dog with a snipy head, long legs, and often a tail which curls over its back. Although they have the ability to carry out whatever task is required of them they are designed for speed — to whizz after a retrieve and catch the eye of the trialing judge. It is unfortunate that these dogs have lost many of the physical characteristics which were so desirable in their ancestors.

There are few sportsmen of the 1980s who participate in moorland shooting, which has resulted in the steady decline of the pointer and setter breeds as working dogs. In their heyday these breeds were used in conjunction with falcons and usually worked in pairs. The most classic pointing dog is the English pointer which, unlike other breeds of gundog, is happy to work for anyone, not just its master. The setters, Irish and Gordon, are rarely seen as working dogs, though their sleek elegance has made them popular as pets.

One kind of dog which has become fairly popular in Britain over the last few years is the all-rounder which will point, flush, and retrieve. These dogs were developed by Continentals and certainly do have their uses, the most popular breeds being the German short-haired pointer (the GSP), the weimaraner, and the Hungarian vizsla. The GSP is the most popular of the three. I feel, however, that these all-round dogs require more attention than the average sportsman can give. They have the ability to work most other breeds into the ground but are so active you must pay a great deal of attention to them as they race backward and forward at a speed which makes springers look positively slow. They are air scenters and have excellent noses, though because of their thin coats they are not best suited for cold water or rough cover. They can be superb dogs to shoot over as they point at game, before flushing on command. However, they are to my mind more suited to the man who is more interested in working his dog than shooting.

Having decided which type of dog most suits your needs you must now plan your timing. It is better to have a puppy in the early spring as this gives you the summer months during which time you can bring the dog up through its most vulnerable period. It is also vital that you realise from the outset that your dog must not be used until at least his second season, no matter how well-trained it appears to be. Premature exposure of any young dog to the shooting field can only bring disaster, creating undesirable faults, and makes the animals unsteady, opening the door to a whole new area of unnecessary problems.

When buying a puppy for training as a gundog one of the most important points is quality in the pedigree. Make sure the dog is of good working stock, a good indicator being the number of field trial champions and winners in the pedigree. I would certainly suggest that you avoid any dog that is not of a proven working strain, as it will almost certainly end up a waste of time. It makes much more sense to start off with the right basic material since no matter how pretty the sow's ear you will never achieve a silk purse.

Inherited quality plays a vital part in the purchase of your dog. You should be able to assess the puppy's potential from the last three generations on the pedigree. Some authorities believe that show winners in a pedigree (denoted by the letters CW or Sh Ch) are undesirable and should be avoided. I believe that you are quite safe if your puppy has show blood in it no closer than great-grandparents. Any show blood in his ancestry will give the litter a more 'true-breed' appearance, and it is quite possible to achieve good looks and brains in the same dog.

Choosing a puppy from a litter of six or seven is not easy but I believe that, along with other qualities which I will mention later, you should choose the puppy whose looks are most appealing to you. I find it difficult to gain any rapport or communication with a dog I consider to be unattractive. Along with good looks I look for a young puppy which will quickly recover after a sudden fright, and shows boundless energy. Other good qualities are boldness and friendliness. Since your dog will have to work continuously with humans the friendliness of a puppy toward people is very important. Check that the puppy's limbs are straight and strong, that the skull is well shaped and the eyes friendly and dark brown. The upper jaw should fit neatly over the lower and altogether the dog should look healthy.

Having chosen your puppy ask the vet at what age it should be inoculated (generally after 14 weeks). This treatment is essential. It is also important that you prepare the ground before you bring the puppy home. This means that if you are intending to keep it outdoors in a kennel you should have this ready for it to take up residence straight after purchase. It can be unsettling for a young puppy to be kept inside your house when it is first brought home, and then moved out to a kennel a week later.

Your kennel should be dry and completely free of draughts. It should have an outdoor run with an elevated platform for the puppy to lie on outdoors. The sleeping section should also be raised off the floor with plenty of warm bedding such as straw, which must be changed frequently. If you want to make your training easier and have a 'one-man' dog, then you should endeavour to keep your dog in a kennel environment, away from the distractions of the home and your family. It means that from the outset the dog will be used only to you, and will ignore others.

However, if you cannot have, or do not wish to have an outdoor kennel, and are keeping your dog indoors, provide a special bed-place from the start — a warm corner that the dog will know as its own, that is a retreat free from disturbance. You will help to make the puppy feel secure in its new home. Most six-week-old pups will cry when you put them to bed at night and they are left on their own. If your dog does this do not go to it for all it wants is attention and once you start answering such calls it will soon learn that yodelling brings results, and you will discover that you have given yourself another set of problems.

A good motto to work to on the subject of diet is that what you put into your dog governs directly what the dog will become. Therefore one must take every opportunity to give the dog protein and nourishment. As a general rule, from the time a puppy is weaned at about six weeks until it is three months old it should have four meals a day; from three to six months three meals; and from six to nine months two meals a day. After your dog is nine months old it should do perfectly well on one meal a day. Like people individual puppies may require more to eat than others.

If your dog appears to be losing weight check for worms and de-worm it with tablets available from any vet. If there is no sign of worms then feed it a little more. If it is gaining too much weight cut down his cereal intake. A good feeding schedule for your puppy is:

Morning meal — puppy meal and milk

Afternoon meal — chopped meat and puppy meal, plus a few drops of liver oil or other vitamin/mineral supplement

Tea-time — as afternoon meal, without supplements

Evening meal — as morning meal

I am a great believer in adding raw eggs to my dog's evening meal, and from six months on a daily egg will do nothing but good.

Your very small puppy will find its new world, away from mother, brothers, and sisters, a bewildering and sometimes frightening place. Like a young child it will need to explore its new environment and should be only gently chastised for wrongdoings. Allow it to be a cuddly puppy for the first couple of months — playing with the children, giving it a few toys of its own, will do no harm. However, your family should also be made aware that the puppy must not regard everything on two legs as a source of fun and games, and visitors to your home should be asked not to fuss over your new pup, difficult as that may be. While I preach the necessity of showing your dog love and affection at an early age, which it will readily respond to, it must not at any time get the idea that all people represent petting. We have all seen the happy picture of a welcoming young puppy jumping all over people with enthusiasm. This is to be discouraged as there is nothing more frustrating than owning a dog which regards every human in sight as a potential source of fun, and anyone who has ever received the attentions of an over-friendly muddy adult dog will know exactly what I mean.

It is wise to remember than an untrained or badly trained dog is not just an inconvenience to the owner; it can also be a nuisance and a hazard to other guns. So strive for field trial standards in your dog — the basic requirements for rough shooting are after all the same — steadiness, discipline, and the ability to find and retrieve game. The little extra time and patience you expend to achieve this can make the difference between a frustrating day's shooting with your dog and an enjoyable one.

There are many important factors involved in the correct training of a gundog, but none is so crucial as patience on the part of the trainer. It is never permissible to lose your temper and act irrationally with a dog. If you are getting frustrated return the dog to its kennel immediately — until both of you are ready for another try. It is also very important that throughout the training you avoid boredom and repetition. Training should always be fun for both of you, and I believe that if you make the tasks simple and enjoyable you cannot start the puppy too young.

The first part of the training is concerned with basic discipline, which means that you are in charge of your dog at all times. Most bad habits found in working dogs are the result of lack of control at an early age.

As soon as you acquire your puppy give it a name and use this at every opportunity, particularly at feeding times. If the name is associated with food the dog should quickly recognise the sound and respond to it without hesitation. Another lesson you can teach your dog when still quite young is to sit on command, and this is also performed at feeding times. Place the food above the dog's eye-line (which will throw its weight back into a semi-sitting position), put one hand under its chin, and with the other hand gently push the hindquarters down into a sitting position. While doing this repeat the word 'hup', the command to sit. Then put the food down for it to eat. By repeating this exercise over a week or so your puppy will learn to sit quickly and painlessly.

When the dog sits on command, without having its hindquarters pushed down, you can start using the hand signal in conjunction with the word command. As you say 'hup' raise your right arm above your head. Get the dog used to both commands delivered together before using the visual one only.

Returning for a moment to the very first lesson your dog learned — its name — you will see that even at this early stage you are beginning to teach it to come to you on command. As you will be using its name as often as possible you will also be calling it to you many times. When it comes give it encouraging pats. You can start using the visual command to come from an early age. When you call your dog's name, pat the side of your right thigh with your right hand, at the same time keeping your arm stiff. It should be a clear, almost exaggerated signal which the dog will be able to recognise at a distance. It will soon associate both signals with coming to you and receiving a rewarding pat.

All the exercises or lessons you have given your young dog before the age of six months are really good manners rather than training, and are the basic disciplines that all dogs should learn, pets as well as working gundogs. You should not try to introduce your puppy to gundog training before the age of six months, as up until that time it is really too young and 'puppyish' to be able to take it all in. Throwing a ball or a simple decoy for your four-month-old pup to pick up is not teaching it to retrieve; it is simply encouraging it to enjoy performing 'fetch' games for you and planting the seed of enjoyable retrieves in later years.

Dog Training Equipment

You will need a few pieces of equipment if you are to train your dog properly. Some will not be required until your dog reaches an advanced stage in its training, and others will be needed from the beginning. They are as follows:

Whistles You will need two whistles, of differing tones. One whistle, which you will use for close work with your dog, should have a muted tone. The other, which should be used (later in your dog's training) for distant work, should be loud and piercing, enabling the dog to hear it when far away from you. Avoid multiple-pitch whistles, and those of such a high pitch that only your dog will be able to hear.

Dummies or Decoys Two dummies of different weights is the minimum number you will require, and if you are making them yourself you might as well produce three or four at the same time, for use in your dog's advanced training work. An empty washing-up liquid container makes an excellent decoy when filled with sand to give weight, cushioned with rags to make it soft, and stuffed inside an old sock to hold the thing together. The amount of sand you put inside the bottle will govern the weight of the dummy. Do not use wood nailed with rags. Wood is hard, with little 'give', which can encourage the dog to hold it too hard. Sharp objects such as nails and tacks can also work loose, giving your dog a nasty mouthful when it goes to pick up the wooden dummy. You can also use manufactured decoys, available from all good field sports suppliers and gun shops.

Fur and Feather Dummies Using either your washing-up liquid bottle dummies, or bought ones, simply tie a pair of bird's wings, pheasant or duck, to the outside making sure that they are secure and will not flap about as your dog carries the dummy in its mouth. For a fur dummy use a rabbit skin bound securely to the dummy. You must try to imitate the weight and feel of live game before the dog moves on to the real thing.

A Blank-firing Starting Pistol Available from all good sports shops.

A Dummy Launcher Available from gun shops, or look up the advertising sections in the shooting press.

From Six Months Old

At the age of six or seven months you can think of teaching your dog the rudiments of gundog training, and the first lesson to give it is to respond to the whistle. When you call the dog, using its name, give two short toots on your dog whistle immediately after. You can combine the visual hand signal to come to you with the whistle signal, and the dog will soon understand that the two-toot signal on the whistle means exactly the same as the calling of its name or the hand signal.

When you are absolutely certain that your dog understands and obeys the double toot whistle command, you can teach it the whistle command to sit. As you raise your right arm and say 'hup' give a single toot on the whistle. It should be a two-beat toot, as the shorter one-beat toot is a different command, which you will teach the dog later in its training. As with the whistle command to come, combining the whistle command to sit with either or both of the other known commands (oral and visual) will soon teach your dog to respond as easily to the whistle.

Your dog will now sit to word, whistle, and visual command. It should stay in position until told to move, and will respond and come

A Fen trap: set *(right)* and closed *(left)*

English springer spaniel puppies; the best choice of dog for the rough shooter

A young spaniel learns control

The labrador, best of all dogs for water work

to its name, whistle and hand signal. It is most important that your dog can do these individual tasks perfectly before you combine them.

Take your dog to a field you can use for training. A public park is not a good place since there will be too many distractions there for a young, inexperienced dog. If your field has rabbits walk them off before entering with your dog. There can be nothing more upsetting to your dog's training than hordes of rabbits fleeing in all directions as you put it through its paces.

After you have let your dog have a run to empty itself and get rid of excess energy, make it sit. Move a little distance away and call it to you. If it gets up and runs after you before being told to do so, firmly take it back to its original position and repeat the process. Gradually move further and further away from it before calling it to you or walking back to it. As you should never let your dog anticipate your next move, alternate calling it to you with you walking back to it, and give a rewarding pat.

Take time over each lesson you teach your dog, and don't make do with shoddy responses from your dog in your haste to get on to the next stage. Be a firm schoolmaster and allow your dog no quarter in its response to your commands.

It will be of great benefit at this stage if you introduce your dog to the car. Start by putting it in the vehicle for a few moments before doing any driving. This will get it used to the strange smell. Then gradually take it on longer and longer journeys. Leave it inside on its own for a short while. Gundogs have to spend quite some considerable time in a car, so the sooner yours gets used to travelling and settling down in one the better.

Your dog should be a pleasure to take out with you by now, and ready to get down to the heart of its training — the finding, retrieving and delivering of game. You will find that once you have reached this stage you should have a better understanding of your dog's temperament and character and will be able to take great pride and pleasure in the next part of the training.

Retrieving

Use a bought decoy (Turner-Richards have a good selection of purpose-made decoys) or a home made one, using an old washing-up liquid bottle filled with sand and bound up with old socks or rags. When you introduce the dog to the decoy make it sit and let it have a good smell of it. Then throw the decoy a short distance away, allowing the dog to watch. Encourage the dog to go for the decoy with the word 'fetch'. It will quickly discover that retrieving is great fun. However, this should always be a controlled exercise. Never allow your dog to run willy-nilly as you toss sticks or decoys in all directions for it to chase after.

Before you throw the decoy make your dog sit and keep him in the sitting position for a few seconds before sending it out with the word 'fetch'. As it nears the decoy repeat the word 'steady' once or twice to let it know that it is getting close. When the dog reaches the decoy pat your right thigh and encourage it, calling its name and giving two toots on your whistle. As it brings the decoy in, raise your right hand to give the sitting command while holding it gently under the chin with the other hand. This should have the effect of the dog holding the 'game' high. Now gently take the decoy out of the mouth, saying the word 'dead'.

Normally this exercise can be achieved easily. However, some young dogs get the idea that they wish to add their own variation to the game. One such can involve running around you and keeping the decoy for themselves. If your dog starts this immediately walk backwards encouraging it on, and when it has come up to you make it sit, and then remove the decoy from its mouth. If it persists in its reluctance to return to you with the decoy, forget it for that day and return home. To continue would serve no purpose and could only make you frustrated and the dog bored.

Once your dog has started retrieving it is time to introduce it to water. Take it to a shallow pond and encourage it to paddle. If you don your wellingtons or waders and go in yourself it will be more keen to go into deeper water with you. Give it a short retrieve in the water, throwing the decoy progressively further and into deeper water as the dog gains confidence. Remember that although some dogs will plunge in from the beginning without hesitation, water is a new environment and even the best labrador must learn to develop its swimming skills in a variety of potentially frightening situations — weeds tangling its legs, cold water, water in the ears, and so on. If your particular dog is persistently reluctant to swim, on no account push it, throw it or force it to water. The dog is reluctant because it is afraid, and patience and slow perseverance are what will eventually overcome this fear. Retrieving from water uses the same basic principles and methods as retrieving on land, the only difference being that your dog will be swimming and not running to get the decoy.

To prevent boredom in the formative training period try never to give your dog more than two or three retrieves per session. Do not allow it to retrieve every decoy you throw. In at least one out of three retrieves pick up the decoy yourself, leaving your dog in a sitting position. In this way you will avoid the dog anticipating the command to fetch, and this should be continued throughout its working life. It does no harm at all to the most experienced working dog to be left sitting while its owner picks up the occasional bird. Every effort should be made to prevent the horrors of running in, since once an animal does this it is exceedingly difficult to overcome the problem.

Quartering

Hunting dogs such as spaniels and pointers must learn to quarter ground as they range in front of the guns searching and flushing out game. You can also teach your labrador to quarter, unless it is your intention to use it for trialing, in which case it is better not to do so. Hunting dogs will quarter ground naturally, but they must be kept in check, their abilities developed and adapted to suit you, and the speed you like to walk at.

Use an area of ground that has ground game, and walk them off. Take your dog to the downwind side of the field and start walking forward slowly into the wind. Tell your dog to 'get on', in front of you, indicting this with your arm. When your dog is about 20 yards from you give a short (one beat) toot on your whistle, and call its name. It will start to come to you. As it does so wave your arm to the left (or right, it does not matter) and walk that way. The dog will run in that direction and will pass you. As soon as it does this give it the turn signal (one short toot) once again and walk to the right, indicating with your hand.

Walk across the field in this zig-zag manner, always giving the 'turn' toot on the whistle, calling the dog's name at the same time, and showing the direction you are about to take with a hand signal. Make a fuss of your dog when you have reached the end of your training ground, and repeat the lesson, starting from the same point and working into the wind. You should soon be able to walk in a straight line behind the dog while it does the zig-zagging, being turned with the occasional whistle from you when it ventures too far to one side.

Scenting conditions will play a major part in your dog's performance at quartering. On some days it will seem to be doing well, yet on others it will miss everything. Many factors play their part in creating good or bad scenting conditions, but the weather probably more than most. As a generalisation, when it is very hot, very cold, dry, or extremely wet, the scent can dissipate rapidly, and the best time to work a young dog is in the early morning before the ground has dried out.

Advanced Training

The next stage is to introduce the young dog to the sound of gunshot. Some people may think it unnecessary to approach the introduction with such care, but dogs can be frightened by the first gunshot they hear, if it is too near to them. This can cause them to flinch at the sound of a gun thereafter. So it is just not worth the risk.

The best way to do this is with the help of a friend. Take your dog into your training field and ask your friend to walk about 100 yards away. Make your dog sit and squat down beside it to comfort it. At your signal your friend should fire a starting pistol. The dog will be comforted by your presence and take little notice of the loud bang. If this is the case ask your friend to walk toward you and at 20 yards intervals stop and fire the pistol again. If at any time your dog shows worry or concern, ask your friend to stop. Go on to something else and return to the pistol exercise another time.

Your dog should be gaining in confidence and ability all the time. It will now be about nine or ten months old. However, if you started training later, or are takings things slowly and easily, it may be well over a year before you get to this stage. Whatever you do, do not rush your dog. Always master one lesson before moving on to the next, and if the dog seems to be taking its time, so be it; you will have to slow your training schedule accordingly.

Retrieving should now be rock steady. You should be able to stand well back from the dog, throw the decoy, and the dog should not move until told to do so. It should then go straight to the decoy, pick it up and deliver it to you with ease. It should also obey your commands, whether you are standing next to it or 50 yards away, and you should both enjoy your training sessions.

When you know that your dog is completely unconcerned at the sound of the starting pistol fired beside him, you can move on to making it drop to shot. As your dog is working in front of you give it the whistle command to hup and immediately after fire the starting pistol. The dog should sit at the sound of the whistle, and as with all other commands it is thereafter a case of practising blowing the whistle and firing the shot to achieve the same result. It should not take long before the dog will automatically sit when it hears the sound of the shot. As I have said before, however, be quite sure that your dog completely understands the spoken command before you go on to the arm movement, the arm movement before moving on to the whistle, and the whistle before tackling the gunshot. Unless it has mastered each individual command perfectly it will become utterly confused when you move on to the next one.

A dog should never link the act of retrieving with the sound of a shot. Each individual command should be a separate entity. To a dog the bang of a gun should represent the command to sit. Being sent out on a retrieve after that must be an entirely different exercise. Obviously, older gundogs know very well that when their master shoots his gun it usually indicates a retrieve, but that is because they are experienced and have learned what to expect. My own dog for instance gives me a scathing look when I miss, for it will already have marked the bird before I shoot.

To introduce your dog to retrieving fur or feather you should get a second decoy, similar to the one you have been using, and tie either rabbit skin or feathers to it. Take your dog out and firstly throw its regular decoy. When it has been retrieved throw the new decoy out in its place, and send the dog out for it. This should not cause any problems. You can move on to progressively more difficult retrieves with your dog. Throw the decoy into long grass where the dog will have to use its nose, though it should have marked the fall as you threw it. Send it back again if it has come back to you without the decoy. If it is completely lost go and fetch the dummy yourself and give the dog a simpler retrieve to restore confidence.

You can now move on to directional commands using two decoys. Make the dog sit and, standing in front of it, throw one decoy to the left, making sure the dog has marked its fall, and throw the other to the right. Giving the command to fetch, point with your outstretched arm at the first decoy. The dog's instinct will be to go for the second one. If it does this make it sit and put it out on the first decoy. If it persists in going to the second, take the retrieve, make it fetch the first, move to another part of the field and try again. When it has retrieved the decoys in their correct order make a fuss of it after each retrieve, and gradually increase the distance between the dog and the thrown decoy.

The variations and progressions of this exercise are virtually limit-less. For instance, after making the dog sit beside you, throw your two decoys as far as possible, leave your dog and walk 50 yards away. Call it to you and then send it back for the first decoy, and then the second. Alternatively, throw a decoy away with your dog at heel, walk off for 50 yards and then stop. Standing facing the dog, give it the command 'go back', signalling with a downward motion of the arm from head level in that direction, and send it back for the decoy. In the field the 'go back' command is very useful when your dog is not working far enough out on a blind retrieve. You can make it sit, and with the command 'go back' and the hand signal you should be able to make it go further away from you until it hits the scent of the game, and can work its way back toward you.

Any variations which involve the dog using both its mind and memory are to the good, and by this time both of you should be en-joying the exercise. The rapport you can build up with your dog is very gratifying, culminating in the sort of understanding I have with my dog. Like many experienced working dogs, mine will, if con-fused, stop and look back at me for directional assistance, underlin-ing the contact and team work that both of us have strived to achieve.

Until now your dog will have heard only the sound of a shot as a signal to sit. Its retrieves will have been all hand-thrown, and it can respond to directional hand signals. Now comes the time to put all three together. Try to acquire a dummy launcher and take your dog into a field with fairly short grass. Make it sit, then fire the dummy launcher. Send it out for the retrieve, helping with hand signals if necessary. Once the dog realises that the bang of the dummy launcher represents something flying through the air which it may be asked to retrieve it will soon start to mark them as they fall, and will retrieve the decoy confidently from long grass, shrubs, or wherever it has fallen.

Once the dog is capable and at ease marking falling decoys you can teach it completely 'blind' retrieves. To do this hide a few decoys about in your training field and then take your dog out. Let it work in front of you, into the wind, and when you are nearing a decoy fire your starting pistol. The dog will have seen nothing fall, and will therefore be working completely blind. It will rely on your accurate hand signals for direction. Eventually you should be able to send it on a blind retrieve without even the aid of the shot to indicate what it is after.

It is at this stage in your dog's training that it should be introduced to the real thing, both dead and alive. We start with dead game. As previously described, replace the furred decoy with dead game — a rabbit is probably the most readily available and best animal to use for this purpose. The texture, weight, and smell of the game will be different for your dog, but after only slight hesitation it should take it in its stride.

Repeat your blind retrieve exercise, as described previously, using the dead game. The stronger smell should make retrieving it easier for the dog.

When the dog is accustomed to dead game you can introduce it to live animals. If you have access to a game pen so much the better. If not you may be able to borrow one or two pet rabbits (or even buy them). Put them in a small fenced area and bring your dog to them. Make it sit, retreat out of sight and watch. If the dog makes any attempt to chase stop it immediately. After a few visits to the rabbit pen the dog should realise that chasing is forbidden. I believe this exercise is extremely important. I have seen many working dogs dangerously tied to their masters because of their unsociable habit of chasing after rabbits, hares, and anything else they may take a fancy to.

The basic training of your dog should now be completed. But it still lacks experience so do not expect it to perform perfectly every time. Everything will slot into place after a few shooting seasons. It is very important that you do your utmost to avoid your young dog picking up a struggling runner, since the more it struggles the greater the pressure it must use to hold it and, hey presto, your beautifully trained dog develops a hard mouth! During its working life the dog is going to encounter many runners and other live game which it should retrieve perfectly well once it has the experience of knowing a live retrieve from a dead one.

Remember at all times that each dog is an individual, just like its owner. Be aware of your animals's character and, using a combination of both my advice and your own common sense, you should be able to solve any faults that may arise.

White hare

6 THE RIFLE

Ever since the invention of firearms men have strived to hit a target with increasing accuracy, force and rapidity, and from ever greater distances. Military requirements have always led the field and it is not surprising therefore that the sporting rifles of today derive most of their features from military rifles of the past. However, like the shotgun and indeed the bow and arrow, there has come a point in the history of the equipment where, with the possible exception of some minor technical improvements in material strength, the machine can progress no further. Paradoxically, greater sophistication of engineering and gunsmithing skills has led to greater simplicity and refinement of the end product.

With the stalking rifle, this high point was reached in the late nineteenth century when the brothers Mauser came, after long deliberation, to the conclusion that the very best way of taking a rifle cartridge out of a box and closing it securely inside the back end of a barrel, so that when fired its resulting gases blew the bullet out of the muzzle in complete safety to the shooter, was by means of a common door bolt. The firearms historian will rightly point out that Mauser did not invent the bolt-action rifle, but the Mauser model of 1898 was the finest expression of this simple concept and, with practically no changes in design, that model has remained the classic rifle action used both for military and sporting weapons ever since. It is a model of sophisticated simplicity. No part is superfluous. It can be used to handle a vast range of cartridge types, from the little 22/250 up to the giant .458 magnum. It is robust and efficient and, without using any tools, the bolt can be dismantled for cleaning or repairs. This action has been copied and imitated by most of the principal rifle manufacturers of the world and today the majority have made changes in the original design only for reasons of economy or to ease production. None has managed to improve it.

Notwithstanding the genius of Mauser, there are other types of repeating rifle which deserve notice and they fall into three groups. The lever-action rifle has, instead of a bolt handle, a lever combined with the trigger guard which, when pushed downwards, slides the bolt backwards, allowing a cartridge to emerge from a tubular magazine under the barrel. Raising the lever pushes the bolt forward and locks the cartridge in the chamber ready for firing.

The pump action, or slide action works in a very similar way to the lever action but in this case the cartridges are fed into the chamber and ejected after firing by sliding the forward part of the stock back and forth with the left hand. Both pump- and lever-action rifles are used normally for relatively low-powered cartridges since their locking mechanism is not as strong as that of the bolt action.

The fourth type is the automatic, which uses either recoil, or more usually a proportion of the gas produced by the burning powder, to release and actuate the bolt. This is the fastest way to reload a rifle and for this reason the automatic is the type used for most military rifles where the greater weight and relatively inferior accuracy are not as important as firepower.

The bolt-action rifle is considered the most accurate of the four basic types and this, combined with its greater versatility, freedom from mechanical trouble and light weight, makes it the most popular choice for a stalking rifle.

But why a rifle for stalking? The answer lies in the basic search for accuracy at long range. The need to strike the quarry with sufficient force to kill instantly at long range calls for a projectile which can be driven at high speed and with great precision. A cloud of shot pellets scattered out of a smooth-bored gun is effective only at short range and against small game. The ballistics of modern cartridges are dealt with in another chapter and will help to explain the need for different bullets, but the principle behind the rifle as opposed to the shotgun needs to be understood first.

The word 'rifle' comes from the old French verb *rifler* meaning 'to scratch'. The scratching in this case is the spiral grooves scratched or cut inside the barrel which grip the bullet in its accelerating passage up the bore and force it to spin at a rate about 150 times faster than an aeroplane propeller. The rotation stabilises the bullet during its flight by gyroscopic action, ensuring that it travels point forward and resists as much as possible the sideways pressure of the wind, thereby keeping it on a steady course to its target. Without rifling the bullet will wander in a haphazard fashion and be ineffective beyond a very short range, as the soldiers of the eighteenth century knew well, before the introduction of the deadly rifle musket.

Duelling pistols of the early nineteenth century were always smooth-bored to give the contestants a sporting chance of survival. A few rare examples of pairs of duellers are to be found where only one pistol is smooth-bored, the dastardly caddish owner thereof ensuring he always got the rifled one.

Rifling is a study in itself to which many books have been devoted. In the early days of mass production of rifled weapons, almost every manufacturer developed his own patented form of grooving with many a bold claim to superiority over his rivals. Ratchet rifling, square-grooved, V-shaped, segmental, hexagonal, gain-twist, multi-groove, oval-bored, fast and slow, shallow and deep; the list is endless. The method of production was almost always by 'scratch rifling' or the cutting of each individual groove, one at a time, with a toothed cutter, which passed up and down the inside of the barrel, removing a fraction of metal with each pass.

Present-day techniques have improved on this and 'scratch rifling' has almost disappeared from the scene. Three other methods are now used for production of rifled barrels — broaching (usually for short pistol barrels only) where a special tool cuts all the grooves in one pass; button rifling where a hardened, olive-shaped button with rifling grooves cut in it is pushed or pulled through a hole slightly smaller than itself, forcing the metal outwards to form shallow grooves and leaving the higher 'lands' in the places where the button had grooves; and thirdly, the most expensive but best method, swaging, hammering or cold-forging.

This last process, used by most of the leading rifle makers, requires a huge investment in machinery but produces barrels of the highest quality and at great speed. A bar of steel is drilled and the hole polished or honed to a size slightly larger than it will be when rifled. The form of rifling itself is ground onto a tungsten carbide mandrel of exactly the size required for the finished barrel. The barrel blank is then swaged onto the mandrel by a giant rotary hammering machine which reproduces inside the barrel the precise form of rifling on the mandrel and at the same time both hardens the surface of the steel and by compressing the molecules increases its tensile strength. The finished barrel has a greater resistance to wear, therefore a longer useful life than barrels rifled by other methods and also will be stiffer by virtue of the increase in tensile strength, therefore more accurate, since it will vibrate less under the impact of cartridge explosion.

In the quest for accuracy every aspect of a rifle's construction must be considered. A rifle is like a tuning fork which when struck by the explosion of the powder, vibrates throughout its length. The stock, the handle by which the rifle is held, is an important part of that tuning fork and must, to ring true, be attached or 'bedded' to the metalwork so firmly that the resonance is the same every time it is struck. If it is not, if the slightest shift of metal in wood takes place, the resonance of the barrel will change from shot to shot and the attitude of the barrel at the moment the bullet leaves the muzzle will vary microscopically. This tiny variation will be magnified enormously during the bullet's flight to a target hundreds of yards away and result in loss of accuracy.

The quality of the wood from which a stock is made can affect the performance of a rifle. The ideal must be a compromise between strength and portability and it has been found that well-seasoned walnut offers the best of both worlds. Walnut is close-grained but not too heavy, does not crack or warp easily and can be readily worked. Some examples of heart wood have an attractive grain and colour and add enormously to the very important aesthetic appeal of a good rifle. Other woods such as beech and maple are also suitable but do not have the beauty of walnut.

This chapter deals principally with the essentials of a stalking rifle which, for reasons that will become clear when we discuss cartridges, means a rifle which fires a powerful centre-fire cartridge suitable for long-range shooting at large game. There is another family of .22 calibre rimfire rifles, for target shooting and vermin control, to which the same basic principles apply but to which, because of their much lower power and more limited range, less attention is usually given. They are, however, the first, or more often after an air rifle the second, step which a shooter normally takes in his or her progress to a high-power stalking rifle.

The maximum effective range of a .22 calibre rifle is about 100yd, although the bullet can easily travel a mile or more if shot at 45°. This must always be borne in mind when shooting and the choice of backstop to any target is of the utmost importance. The rules of safe shooting are simple but easily overlooked.

It is extremely easy, if you are not using your head, to have an accident, and two examples I have witnessed are the following: on one occasion a friend of mine hopped out of his Range Rover and, using the bonnet just in front of the windscreen as a rest, crouched over and carefully took aim with his .22 at some rabbits on a slope about 50 yards away. When he squeezed the trigger not only did he miss the rabbits, but he was aware of a strange sound which he couldn't explain. Only when he stood up did he realise that although he could see the rabbits clearly through his telescope, the muzzle of his rifle had been pointing straight into the slightly raised hinge of his bonnet, and the bullet had gone inside the hinge, knocking the heavy hinge-pin almost all the way out. Where the bullet bounced to no one knows, but it must have nearly parted his hair.

On another occasion when stalking with a friend we crawled into an old sheep fank and took cover behind the wall. The big stag my chum intended to shoot was no more than 100 yards away. The shooter carefully raised himself up, took aim using the top of the wall as a rest, and squeezed the trigger. At the sound of the shot the stag jumped, but neither of us had heard the bullet striking the beast, nor did the animal give any sign of injury. Quickly my friend worked the bolt and took aim again. Just as he was about to shoot I, peering over his shoulder, realised what was happening and sharply told him to stop. The stag ran off as my chum turned on me as though I had had a fit. What he had not seen when he took aim at the stag was a long point of rock sticking out of the dry stone dyke directly in front of his muzzle, and no more than 20ft away. Smack in the centre of it was the round mark where his bullet had struck. So not only must you be extremely cautious as to where your bullet may go, for it may pass through an animal or miss where there is no adequate backstop of a slope or hill, but also at close range using a telescope always check that there are no obstacles in front of your muzzle.

No matter how accurately a rifle shoots, it needs sights with which to aim it. These may be of three types: open sights — a simple blade or post at the muzzle with a notch or V-sight towards the chamber end of the barrel which are lined up with one another and which in turn are aimed at the target; aperture or 'peep' sights consisting of a small ring rear sight set close to the eye on the action with a conventional front sight; and thirdly a telescopic sight. The latter is the most efficient and easiest to use as well as being the most sensitive to damage. For this reason a sporting rifle is usually fitted both with a scope and open sights which can be used in an emergency if the scope is damaged or obscured by driving rain.

To understand the problems in sighting a rifle you have only to appreciate that the eye cannot focus an object near to the eye and another at a distance. One or other will be blurred. With open sights it may be possible to have both the front sight and the target in focus but most people find this practically impossible and the rear sight will inevitably be a blur.

An aperture rear sight, working somewhat like a pinhole camera, helps to sharpen the focus over a much deeper field and the eye will automatically be centred in the hole, but if it is small enough to give sharp focus the light loss will make the target more difficult to see, especially at dusk or on a dark day.

A telescopic sight overcomes these problems with very little light loss and, in addition to putting both the target and the rifle sight on the same focal plane, magnifies the target, making it much easier to aim at the vital area. Scopes come in a variety of magnifications, 4-power being the most commonly used. At this magnification the target is sufficiently enlarged — seemingly only ¼ of the actual distance away — to be clearly seen. Of course, the higher the power the more obvious and distracting are the wobbles of the rifle as you take aim. It is very disconcerting to see your target dancing about in the sight picture and only the steadiest shooter will be able to manage a magnification greater than x6, especially when the adrenalin is flowing fast.

A word about variable magnification scopes: these work like a zoom lens on a camera and have their drawbacks as well as advantages. The range of magnifications they cover may be x1¾-4, x2-7,

Deer seen through a telescope: 1) a correct view 2) an incorrect view – greater magnification gives a misleading impression of the target

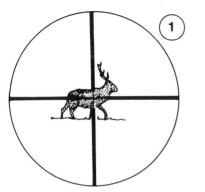
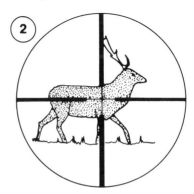

x3-9, and even as high as x4-12. It may seem very convenient to be able to wind up the magnification and pull the target into more visible range but the great danger in doing so is that the shooter gets a totally false idea of how far away it really is and is more likely to wound an animal by hitting a non-vital area.

The stalker must learn the capability of his rifle and how to judge shootable distances for different types of game. The experienced shot will use the familiar sight picture through his scope to assess with great precision the range of his target. Almost without thinking he will adjust his aim to compensate for the trajectory of the bullet, but if he keeps changing the magnification to make the target look bigger this ability to judge distance will inevitably be impaired and the risk of leaving a wounded animal is too great a price to pay. Even vari-power scopes with built-in trajectory-compensating devices cannot replace the shooter's ability to judge the range for himself. Range-finding systems are not foolproof and the time taken fiddling with dials can make the difference between a successful shot and a failure. Stick to a fixed magnification scope, learn to judge distances and keep the equipment as simple as possible — there will be less to go wrong.

Zeroing a Rifle

A rifle which is not zeroed correctly, shooting exactly where you aim it, is as much use to you as a bent barrel. When you are confident that you can place bullets exactly where required it will greatly reduce any misgivings you may have about shooting at an animal. No excuses should ever be countenanced — for you to say your rifle is shooting a little high, low, left or right is an admission of your own laziness or lack of ability to do the job correctly.

Zeroing a .22

You must first take a piece of plain white paper and draw a ¼in wide 2ft long black vertical line down the centre. Pin the paper to a sheet of board (thick cardboard will do) and stand it firmly against a solid background such as a hillside or sandbank, for reasons already explained. Get yourself in a comfortable position, preferably prone. You need go no further than 15 yards from the target where you will firstly align your shots on the vertical line. At this juncture range is unimportant, but accuracy on a vertical is paramount. If you already have your scope partially zeroed it may not be necessary to use a paper as large as 2ft, but if you are starting from scratch it certainly is.

Fire your bullet. At that range you will be able to see clearly where it has struck the target, and adjust your sight reticules accordingly. It should not take you long to bring the bullet on to the vertical line. With a red pencil you should then mark a small circle around each bullet hole. Now move back to 50 yards from the target and, getting yourself in a comfortable, prone position, fire three bullets, and once again check them for vertical accuracy.

When you are certain that the vertical is correct draw a ¼in wide horizontal line on your target, making a cross. From 60 yards lie down and fire one shot at the target. Go to the target and check where the bullet has struck, marking it with a small ring. Repeat this exercise, adjusting the reticules as necessary until you are firing every bullet on the centre of the cross. Some people feel happier taking two shots between each adjustment of the reticules to eliminate possible shooting error. It really is easy to fine zero your sights on the cross.

When you are certain you are zeroing neatly then it is time to use a third target. Get another piece of white paper (it need not be as large as 2ft for this exercise) and draw a heavy black dot, about 1½in in diameter. At 60 yards you should be able to consistently place all bullets in the spot.

Zeroing a Full-bore

Modern full-bore rifles are particularly flat on trajectory and the correct way to zero one with a telescope is to start with a plain sheet of white paper with a 2ft long vertical line, as described for zeroing a .22 rifle. I cannot stress too strongly the importance of never putting more than one bullet at a time in your rifle when you are zeroing it.

Lie prone, and at a distance of 50 yards fire one bullet at the vertical line. Go to the target, check where the bullet has struck, and mark it with a pen ring. Adjust your vertical reticules as necessary and repeat, taking plenty of time. Make sure your breathing is relaxed, and do not hold your breath when shooting. You should squeeze the trigger on a gentle exhalation.

When your bullets are striking the vertical line set a cross target and from a distance of 100 yards align your horizontal reticules. Your intention should be that, although you are aiming at the centre of the cross at 100 yards, your bullet should be striking 1in high. This will mean that at 200 yards your bullet will be 1in low, and that within a range of 200 yards your accuracy on a 3in circle is assured.

103

When your rifle is accurately set a good exercise to practise is to lie in a comfortable, prone position and, putting three bullets in your rifle, aim at your target. Then smoothly and rhythmically fire all three. Many good shots in a stalking situation tie themselves in knots when things go wrong, fumbling in panic as they try to reload quickly.

If you can afford it then a worthwhile piece of additional equipment for a scope mounted rifle is a scope aligner. This small optical device comes with a range of barrel inserts to cover all calibres, and is simplicity itself to use. Fit the aligner to the end of your barrel, then simply adjust the reticules to the range you wish your rifle zeroed. This instrument saves expense and inconvenience, and is excellent, though I would advise that, having set your scope, you always check one or two bullets on the target for certainty.

Shooting Positions
It is not always possible to shoot from a prone position and the competent rifleman should be able to handle his weapon in several positions. One is to sit with your feet apart and your knees drawn up. Place your elbows against the inside of your knees. With a gentle outward pressure of your elbows you will create a stable shooting position. Another position is on one knee with the other leg bent in front, your weight leaning back on the kneeling leg, and your buttock on your heel. With your elbow on your raised knee you will find this an excellent and stable shooting position.

The Rifle Cartridge

The range of rifle cartridges for target and game shooting is enormous. One of the ways in which equipment manufacturers stimulate interest in their products is to develop something new and hopefully better for some particular purpose. Nowhere is this more true than in rifle calibres and bullets. The fact is, however, that for all the types of larger game in the British Isles only two calibres will suffice, .243 and .270. Between these two the choice is largely a matter of personal preference, as later ballistic comparison will show.

The choice of calibre will depend on the type of shooting you have in mind. Most of this chapter will be concerned with centre-fire high-power cartridges for deer stalking but it should not be forgotten that most rifle shooting in this country is with a .22 rimfire which, though illegal on deer, is nevertheless potent enough to despatch the largest

A good bag of partridge

A Parker-Hale stalking rifle, with Pecar scope

Parker-Hale British series stalking rifles: 1200H *(top)*; 1100H *(centre)*; and 100H *(bottom)*

A red deer calf

A rare, almost white, red deer calf

species if well placed. The smallest calibre permitted for use against deer in England and Wales is .243 although in Scotland the slightly smaller .222 is allowed. This regulation is designed to ensure that game will be shot only with a bullet which delivers sufficient energy to produce a clean kill in a vital area.

The expressions centre-fire and rimfire refer to the method of ignition of the powder inside the cartridge case. The very small .22 cartridge (.22in being the diameter of the bullet) can be made with a thinly drawn brass case since it does not need to contain a very high pressure of gas. The easiest method of ignition for .22 rimfire is to pack the priming compound, like the material used for toy cap guns, in the hollow rim of a case, which is thin enough to collapse under a blow from the firing pin of the rifle. When the rim is pinched the priming compound explodes and ignites the slower burning powder which produces a fast-expanding mass of gases. Having nowhere to go, they push the bullet in front of them up the barrel and explode out of the muzzle producing the crack which we all associate with rifle fire.

The need to contain the very high pressure developed by a large quantity of powder led to the design of the centre-fire cartridge. This, as its name implies, is ignited by a cap with its own self-contained anvil in the centre of the cartridge base. The base of the cartridge case is its thickest part since it has to perform three functions: to contain the primer or cap; to form a strong gas seal; and to provide a means of extracting the empty case from the chamber after firing. The extractor is a claw which hooks into a groove cut round the cartridge base. It has to overcome considerable pressure after firing since the explosion expands the cartridge case into the chamber with a force of around 50,000lb per square inch.

The bullet from a .243 cartridge leaves the barrel at about 3,000ft per second. The plain lead bullet of a .22 rimfire cartridge which is travelling at only one third of this speed would melt under the friction generated if pushed at the speed of a .243. To protect the high velocity bullet in its passage up the barrel, a jacket of gliding metal is used around a lead core.

The most efficient way to transmit the striking energy of the bullet to the target is to ensure that it stops in the shortest possible distance. This is achieved by constructing a bullet with a soft nose which will spread into a mushroom shape on impact. The jacket is usually left open at the point with the lead core exposed. This soft-point bullet is the most common type in use for game shooting. The hollow point bullet is very similar to the soft point but has the hard outer jacket extending beyond the lead core leaving a hollow which promotes expansion but will not be deformed in the magazine of the rifle when it recoils. Full jacket bullets with a solid point are confined to military and target use being designed for deep penetration, not expansion, and are outlawed for most medium game shooting. In addition to these three basic types, ammunition manufacturers have developed a number of other designs to improve stopping power and for special purposes. These include the dual-core with a soft lead core at the front for quick expansion and hardened lead section at the back to increase penetration. The American-designed Nosler bullet has a partition formed by the jacket and a separate lead core on each side of this partition. There are plastic-pointed, silver-tipped and spire-pointed bullets, boat-tailed and flat base, round-nosed and pointed bullets, all with their particular characteristics and advantages.

A pointed bullet is more streamlined and overcomes air resistance better than a round-nosed bullet, but on the other hand the round-nosed type is often steadier in flight and less easily deflected by leaves, grass or twigs. An excellent all-round bullet for the medium-sized game species found in the British Isles is the semi-pointed type, a satisfactory compromise which results in good flight characteristics and expansion on impact.

In choosing the right cartridge for your type of shooting you will have to consider two principal factors — calibre and bullet weight. Let us compare the two most widely used calibres for ballistics. First the .243 Winchester which was developed from the heavier .308 Winchester, more familiar to most people as 7.62 NATO, the standard infantry weapon calibre of the US and NATO forces for the past twenty-five years.

The .308 diameter bullet was considered by many to be too heavy for the smaller species of deer. Too much penetration often resulted in the bullet passing clean through an animal which might well be lost and left wounded on the hill. The heavier the bullet, the slower it will travel unless propelled by a larger powder charge. This will be limited by the capacity of the case and the strength of the rifle action. By reducing the neck diameter of the .308 case to .243 in a smaller, lighter bullet could be propelled at a higher velocity, thereby giving a flat trajectory with very little bullet drop over a range in excess of that at which the experienced stalker would normally contemplate a shot. The high velocity of the small bullet produces as much striking energy on the target as a heavier but slower bullet up to a certain range. Beyond a certain range a heavier bullet will retain its velocity and therefore its striking energy better than the lighter one which loses velocity more quickly as air resistance and gravity take over.

The muzzle velocity of a .243 with a 100-grain bullet is 3,000ft per second. Its trajectory curve can be plotted by reference to the residual velocity at certain ranges and the striking energy at those ranges calculated by the following formula:

$$\frac{\text{Velocity squared x bullet weight}}{450250} = \text{Energy}$$

If the velocity is expressed in feet per second and the bullet weight in grains you arrive at a striking energy in foot pounds. The comparison chart for .243 .308 and .270 showing velocities at ranges up to 300 yards and the relative striking energy in foot pounds looks like this:

Calibre Bullet Weight	Velocity (ft per sec)				Energy (foot pounds)			
	Muzzle	100yd	200yd	300yd	Muzzle	100yd	200yd	300yd
.243 Win 100 gr.	3,070	2,790	2,540	2,320	2,090	1,730	1,430	1,190
.270 Win 130 gr.	3,140	2,884	2,639	2,404	2,847	2,401	2,011	1,669
.308 Win 130 gr.	2,900	2,590	2,300	2,030	2,428	1,937	1,527	1,190

Few shots will or should be taken at ranges above 200 yards and the average will probably be nearer 150 yards. At 200 yards the striking energy of these three calibres is quite sufficient to produce a clean kill. The advantage of the smaller calibres, .243 and .270 is that their higher velocity results in a flatter trajectory within normal stalking range so little or no allowance need be made for bullet drop.

The following table shows the trajectory of the 100 grain .243 and 130 grain .270 and .308 bullets, expressed in inches above or below the line of sight (which of course is dead straight and approximately 1½in above the centre of the bore).

Line of sight 1½in above centre of bore
+ indicates point of impact in inches above,
− in inches below, sighting point

	Sight at (yards)	25yd	50yd	100yd	150yd	200yd	300yd
.243 Win							
100 gr.	100	−0.7	−0.2	0	−0.9	−2.9	−10.6
	180	−0.5	+0.3	+1.1	+0.7	−0.7	− 7.4
	200	−0.4	+0.5	+1.4	+1.3	0	− 6.3
.270 Win							
130 gr.	100	−0.8	−0.3	0	−0.8	−2.7	−10.7
	180	−0.5	+0.2	+1.0	+0.7	−0.7	− 7.7
	200	−0.4	+0.4	+1.4	+1.3	0	− 6.6
.308 Win							
130 gr.	100	−0.7	−0.2	0	−1.1	−3.7	−14.2
	180	−0.4	+0.5	+1.4	+1.0	−0.9	−10.0
	200	−0.2	+0.8	+1.9	+1.7	0	− 8.6

If the rifle is zeroed, or sighted-in so that the bullet strikes exactly at the point of aim at 100 yards, a .243 calibre bullet will have dropped 2.9in at 200 yards and a .270 with 130-grain bullet only 2.7in. This is insufficient to carry the bullet outside the vital area and may be disregarded. The .308 calibre bullet of 130 grains will have dropped 3½in at the same range because of its slightly lower muzzle velocity and greater frontal area which creates more air resistance.

Although .243 is considered by many to be the ideal calibre for all species of deer and the larger vermin found in this country, there is a school of thought which favours a margin of additional stopping power which is afforded by the .270. This cartridge is based on the .30/06 Springfield, which was developed in the United States in 1906 for military use and fitted with a .308 calibre bullet. Necked down to .270 its ballistics are superior to the parent cartridge and its lighter, smaller-diameter bullet makes it suitable for a very wide range of game species.

A heavier bullet of 150 grains is more suitable for stalking in woodland where short-range shots are the norm and where there may be leaves or twigs to deflect a lighter bullet. The longer-range shots needed in open country put a premium on high velocity to reduce bullet drop so the 130 grain .270 will usually be the choice of a Highland stalker.

Remember that your rifle will shoot in a different place with different weight bullets and also with cartridges of the same specifica-tion but from different manufacturers. Once your rifle is zeroed you should stick to the same brand and same bullet weight. Every time you change cartridge the rifle will have to be sighted afresh. You will find that a change of bullet weight will not only affect elevation but, because of the slight twisting of the barrel along its axis as the shot is fired, also the lateral placement of the shots. Differences between brands of cartridge loaded with the same weight bullet will come about not only as a result of different powder loads and velocities but because the rate at which the powder burns may also vary and this will affect the barrel vibrations as the shot is fired. If the attitude of the barrel at the moment of departure of the bullet differs even slightly, this will change the trajctory of the shot.

In this chapter we have only been able to look in detail at the most widely used calibres but the choice is enormous. Leading manufac-turers of rifles will offer as many as twenty different calibres in their range, to cater for needs as diverse as those of the African game war-den, the American hunter in the Rockies whose target may rarely be nearer than 400 yards away, or the 'varmint' hunter who shoots small game across the open spaces of the desert. A close study of ballistics provided by the ammunition manufacturers will give you a good idea of the performance of the many calibres available but in the British Isles if you stick to .243 or .270 you will not go far wrong.

Game for the Rifle

In the United Kingdom very few men can justify having more than two rifle calibres in their gun cases — a .22 rimfire, and one of the larger calibres — a .243 or .270. The .22 is an extremely versatile little weapon, and a great deal of pleasure and sport can be gleaned from the sensible and responsible use of it. Rabbits, where they are plentiful, can provide an almost unending series of sporting targets, and with the combination of a quality telescope and silencer, this weapon can be really deadly. However, it is totally unsuitable for shooting anything larger than a hare. Though it is common practice in the more remote areas of the United States to use the .22 for shoot-ing grey squirrels, roosting crows and so on, there are few areas in Britain where this practice would be anything other than extremely dangerous, since the bullet, shot at an angle of 45°, will travel fully a mile.

111

The largest game in this country are the five species of deer, and these animals should be shot with a rifle; by law of a calibre not less than .243 (.222 in Scotland). Indeed the man who owns a full-bore, heavy-calibre weapon, .243 or .270, is restricted to stalking one of the deer family, and even that is slightly misleading since, although there is an open season for the muntjac, at the time of writing this book they are not common enough to be taken into account when discussing stalking. Rifles are also used under special circumstances on seals, when the appropriate government licence has been obtained, and a substantial number of foxes are taken with the rifle, particularly on the large estates and Highland areas.

Therefore the British sportsman or woman who wishes to use a heavy calibre rifle is really limited to the four most abundant species of deer: the delicate and probably most difficult to stalk, the roe; the fallow with its palmated antlers; the sika, or Japanese red deer; and the Scottish red deer — the largest stags can weigh up to 300lb. Quality stalking for any of these deer can be had from private estates and forestry companies throughout the country.

Red Deer *(Cervus elaphus scoticus)*
The red deer is the largest mammal in Britain and can be found in the Lake District, Somerset and Devon, though its true home is the Highlands of Scotland, where they are widely distributed in considerable numbers. There are more red deer in Britain today than there have been for many centuries, with a remarkable increase over the last twenty years. In the 1960s there were about 150,000 red deer but now there are 270,000.

Red deer have had no natural predators since the country's wolf population was destroyed in the 1600s, and the only means of controlling their number has been through shooting. However, many estates are reluctant to decrease their overall red deer population, for fear of losing the revenue they earn from sportsmen eager to shoot stags. Also, it is the hinds which should be culled, but few estates can afford the stalkers to shoot the females during the hind season, October to February (at which time the hills can be inaccessible anyway), and hind stalking is not so appealing to sportsmen as stag stalking.

As previously mentioned, the male of the species is known as a stag, the female as a hind, and the young, normally born singly, as calves. Unlike roe, the red deer travel in herds, which number from as few as a dozen up to several hundred. During most of the year the herds of stags and hinds normally live apart. In the summer months the stags remain at the highest levels, where they can get peace and escape from both the heat and irritating flies. The hinds generally favour a slightly lower range than the stags. I must stress, however, that this is a generalisation as both sexes will move around their habitat as weather and food dictate.

Stag stalking is the most expensive, and traditionally regarded as the most glamorous, event in the sportsman's calendar. Although stags can be stalked legally from 1 July, at that time of year they are usually too high and difficult to get at, the very remoteness of their location making it extremely difficult to bring the carcasses out. So the majority of estates prefer to wait until later in the season when the weather is cooler. This makes stalking more pleasant, and puts the beasts in a geographically less difficult situation.

Undoubtedly the best time to stalk stags is in the period from the end of September through to 20 October (end of their shooting season). During that period the rut takes place, particularly over the last two weeks of the season. By then the red deer stags, pushed on by their natural rutting desires, have broken out of their groups and started to search individually for hinds. There can be few more spectacular times of the year than when the mountains seem to reverberate with the roaring of stags as they call out their challenges.

As the hinds come into season from late September to mid October, the stags, which have until then been living on the high tops, start to break away from each other as they move about in search of the females, becoming increasingly more aggressive toward each other. Once a stag has found a group of hinds it will try to take possession of them, and will hold them together, chasing off rivals until sated. It will then retire as its urges decrease. It is very much a case of the larger stags getting the prize, and it is not uncommon to see a large stag, while its herd of perhaps fifteen hinds grazes contentedly, charge from one end of the herd and back to the other, chasing off the lesser stags which hang about the outside of the herd waiting for an opportunity to cut a hind or two out for themselves. It is a fortunate stalker who witnesses an actual fight between stags, since these are comparatively rare. Disputes normally take the form of much vocal threatening, with smaller stags running out of the way instead of fighting, and when fights do occur it will be between fairly evenly matched stags. Although it looks spectacular, with the beasts charging at each other, and crashing antlers, they seldom do any real damage. Obviously wounds do occur and can be quite serious, from the loss of an eye or a bad stab from a tine, but seldom in the animal kingdom will one member of a species do serious damage to another. The antlers are used for display and as a note of size and rank. Although they can be extremely vicious weapons, real damage is not common.

Tracks and Signs
The red deer track or slot is large and easily recognised, with a much rounder print than that of any other deer. It is approximately 8cm long and up to 7cm wide, with distinct marks of the outside sharp edge of the hoof visible, and the inner surfaces of the toes concave. Although red deer claws will normally be seen only in soft ground, their feet splay easily, particularly when the beast is moving at speed. One confusing factor for the inexperienced tracker is that a smaller animal can leave tracks similar in size to those of a sheep. However, if studied it will be seen that they are of distinctly different shapes. When walking, red deer have excellent, clear prints to either side of the median (central) line. At a trot they are not quite so clear, with one just cutting into the other, and on a much straighter course on the median line. Moving at speed their prints are distinctly in groups of four, with their toes widely splayed, and anything from 2 to 3 yards between one group and another, depending on the size of the animal.

Signs of red deer presence differ in the two habitats where the animals are likely to be found in this country. In afforested regions nipped-off tree shoots and buds, and stripped bark can be observed up to a height of 2 to 3 yards. During winter vegetation will be consumed by the deer, particularly kale which is often grown as sheep grazing. In the mountainous Highland areas signs of red deer would normally be found only on soft paths around wallows, or in peat hags. Their signs on the hill are more difficult to detect since deer eat moss, heather shoots and upland grasses, but it is not really necessary to look for signs in this type of habitat as the animals themselves are spied more easily.

Red deer droppings, or spoor, cannot really be confused with any others, unless again with very small beasts. Stags' droppings are 2-3cm long, pointed at one end and flat at the other. Hinds' are round or oval. Colour varies according to food eaten, from milky chocolate to black. During the rut stags leave softer piles, either a compression of their normal individual droppings, or a pat more similar to a cow's, but no larger than a saucer.

Wallows should not be confused with a peat hag or bank. The wallow has an obvious easy entrance, flattened by the passage of countless hooves. Red deer couches are found only in woodland and are flattened areas with good shelter. On the hill deer do not favour habitual lies, and the springy nature of heather conceals where they have been lying. In the summer you are more likely to find where beasts have been lying, and this is almost always just below a sheltered ridge where they can view the terrain for approaching dangers.

During the rut stags give off a distinctive odour which can be detected not only where they have been lying but also by the educated nose several hundred yards downwind. For example, I was once stalking on the hill with a friend, when a heavy mist descended, and visibility dropped to no more than 20 to 30 feet. We were picking our way along a ridge when I picked up the very distinct smell of a stag. Cautiously we worked our way forward for 200-300 yards and virtually blundered into a huge stag and a small group of hinds. I had followed the beast entirely by scent.

Red deer cast their antlers every year around February and March.

Sika Deer *(Cervus nippon)*

The sika is the second largest deer found in the United Kingdom. The males are known as stags, the females as hinds, and the young as calves. They were originally introduced as captive herds in the late 1800s, and the wild population has grown from escapees. They are now found in pockets throughout the country, from the south of England (particularly in the New Forest), northern England and the Lake District, to Scotland, where they are widespread.

The major difference between sika and their near cousins the red deer is that the sika are not so herd orientated, with the stags generally found living alone, and the hinds in small groups of twos or threes. Even in the breeding season the hinds will sometimes wander off on their own. Sika are also more nocturnal than red deer, with the exception of the Highland areas where they have adopted more of the habits of the red, and can be seen in small groups out on the open hill, feeding during the day. However, when given the opportunity sika prefer thick woodland which gives them cover and shelter.

Sika are similar in appearance to red deer, but slightly smaller, up to about 42in at the shoulder. Their antlers are considerably smaller,

Domestic Cat

Badger (Left hind)

Hedgehog (Leftfore)

Brown Hare (Fore)

Blue Hare (Fore)

Rabbit (Fore)

Grey Squirrel (Fore)

— Hoofprints and tracks —

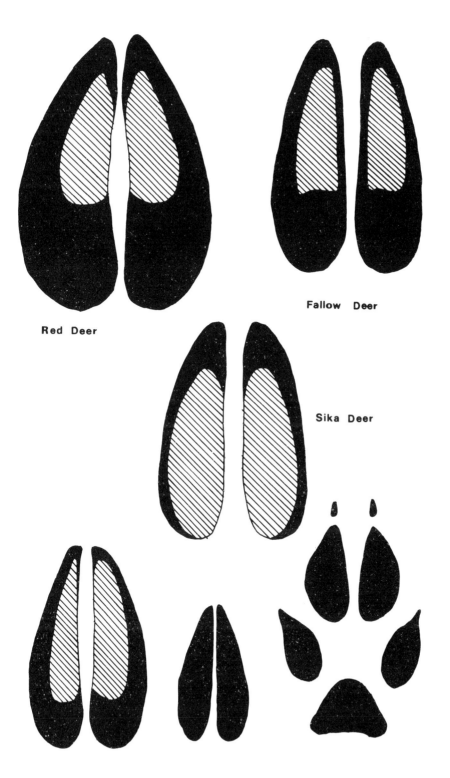

Red Deer

Fallow Deer

Sika Deer

Roe Deer

Muntjac Deer

Fox (Hind)

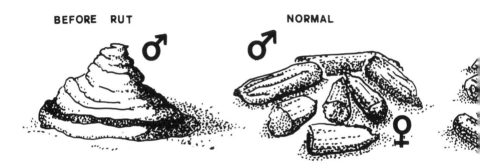

BEFORE RUT   NORMAL

♂   ♂   ♀

Red Deer

Muntjac

Hedgehog   Hare

Droppings

Sika Deer   Roe   Deer

Badger

Rabbit   Squirrel

usually having four points or tines per antler, and no bay tine. The sika has a distinct caudal patch around its rump, and as with roe deer this is flared in alarm. Noticeable if seen from behind are the glands on the outer edges of the hind legs, which are covered in lighter hair.

The species has been known to inter-breed with red deer, though this does not seem to present a problem to either species. During the rut sika stags can be heard to give a cautious whistling sound which can often turn into a high-pitched scream. The hinds make a softer, lower-pitched whistle during the rut. They are reddish-brown, some being spotted — particularly the hinds which can be fairly well speckled.

Since sika generally prefer woodland, the technique of stalking them is as explained Chapter 7. The best times of the day are early morning and early evening.

Tracks and Signs
The novice can easily confuse the tracks of the sika with the red deer, although sika tracks are smaller — 7cm long by 4.5cm wide, with rounded tips and distinct sharp edges. When walking, their hooves are on either side of the median line. Splaying is evident in soft ground, often giving the impression that a larger animal has passed. When trotting, individual tracks disappear and each group of four hoof prints appears on the median line in clusters.

Their droppings are small and pea-shaped. Compared with droppings of other deer sikas seem disproportionately small. They can be oval, round, or oval with a small point at one end, and are most often found in regularly used areas, although the beasts can deposit them when feeding and moving. Apart from the barking of trees and nipped buds and shoots, they do not leave many signs of their presence unless they have access to arable crops when obvious damage may be observed. Their antlers, which are dropped in late March or early April, are more likely to be found in wooded areas where feeding is lush. Couches are found in thick, sheltered brush, though they confusingly use more than one. Their couches are similar to those used by other species of deer, and to be certain you have sika on the ground you must rely on tracks and droppings.

Fallow Deer *(Dama dama)*
The origins of the fallow deer are uncertain since there is no fossil evidence to show it as an indigenous species to Great Britain. Various

theories have been put forward to account for its arrival in Britain, the most likely being that it was introduced by the Romans as a food supply. However, it is known to have been well established in this country by the twelfth century. The males are called bucks, the females does, and the young, which are usually born singly, are called fawns. Fallow deer are very common in most of the southern counties of England, but found only sporadically in Wales. In parts of Scotland they are quite numerous, though not common across the whole country. It is interesting to note that the fallow population living on the islands of Loch Lomond adapted their habits quite happily to swimming from island to island, and it is only in recent years, with the upsurge in tourism and watersports in the Loch Lomond area, that their numbers have declined. An example of their ability to establish themselves firmly in an area is their high population in the afforested areas of central Perthshire.

The fallow has never really developed into a hill animal but has flourished in areas of dense woodland, its preference being deciduous wood, or a mixture of deciduous and coniferous. It seems to have no real attraction for purely coniferous forest. Since fallow rely on camouflage rather than speed to evade predators they favour woodland where bracken is found.

The main distinguishing features of the fallow buck are his palmated antlers and prominent Adam's apple. The species' colouring can range from a brownish-black back with greyish belly to a spectacular light, reddish colour with white spots. The normal fallow has a white rump patch, rimmed in black.

Tracks and Signs
Relative to its bulk, the track or slot of the fallow is fairly delicate, about 6.5cm long by 4cm wide. In soft ground or mud they can be clearly seen, though on harder ground the outer sheath of the hoof, being hard and sharp, often leaves only the silhouette shape of the hoof imprint. The dew claws would normally only leave imprints in extremely soft mud or earth, and when tracking fallow it is unusual to find their tracks as splayed as those of the roe deer.

Their black droppings are found either in a line as the beast has been moving, or in small piles. About 1.5cm long, they are easily distinguished with one end flat and the other pointed if deposited by a doe. Bucks' droppings are also pointed at one end, but the other is dished or concave.

121

If fallow are living in rough woodland areas you should look for signs of grass cropping and tree barking. They will often eat crops if there is arable land nearby, and in the case of root vegetables such as turnips they leave fairly obvious signs of their presence — chewing, extensive damage and of course prints. Due to the rich feeding of the fallow's favoured terrain, their cast antlers are more likely to be found than those of most other species of deer. There is less necessity for the beasts to eat their antlers, which are cast in April and May, since they invariably live in lime-rich countryside, and they get sufficient minerals in their diet.

During the run the fallow bucks will delineate their territories with a distinctive musk, and young trees — most noticeably birches — will often be quite distinctly flayed. Their couches are easily recognised as obviously flattened lying areas in hidden spots, which smell faintly of their sweet musk.

Stalking fallow is best done in the early morning or evening, throughout the season, with equal chance of success. Stalking may be procured from both private estates and forestry companies.

Roe Deer *(Capreolus capreolus)*

The roe is the most beautiful and delicate of the indigenous deer in this country. The male is known as a buck, the female a doe, and the young, which are normally born as twins, as kids or fawns. Roe have been with us since prehistory and have changed little since then. This is evident from remains dug up in various parts of the country. The species has had a fairly chequered career, particularly in England, and was even regarded by some authorities during the early nineteenth century as almost extinct. Although formerly a native of Wales no roe has been reported there for over 300 years. Their absence there is probably due to disappearance of the Welsh forests at around the same time. Roe were never indigenous to Ireland and, although attempts to introduce roe there during the last century appeared to be successful, there are none there today.

Roe have always been disregarded by sportsmen in Britain, with estates keeping numbers in check by organising roe drives. The few individuals who stalked roe in the past were generally regarded as slightly 'cranky'. It has been only in recent years, with the upsurge in interest by Continental stalkers, who have always enjoyed a more enlightened attitude toward this beautiful animal, that estates have viewed the animal with a more respectful eye, albeit tempered with

Ever-alert red deer hinds and calves

A red deer hind nuzzling her calf

A red deer hind; aware of the slightest noise

greed since each one now has a healthy price tag. In a way it is understandable that roe were so disregarded here since most shooters who ever saw one literally tripped over it, blundering upon it by accident. This is entirely different from the true roe stalking situation which requires a great deal of skill, knowledge and stealth, since roe are only really up and about during the hours of dawn and dusk. There can be few more exhilarating experiences for the outdoorman than stalking a roebuck at 5 o'clock on a May or June morning, whether he is using a rifle or a camera.

The roe is a secretive creature and can be found in considerable numbers across most of England, especially the South, and is extremely prolific throughout most of Scotland, particularly in the central Highland region. They can live quietly in an area without the local human population being aware of their presence.

They favour dense woodland areas with a good, clean water supply. When they feed they tend to stick to the edge of woodlands or in rides, though in areas where they are not disturbed they can be found grazing out in the open in the middle of grassy fields.

The summer coat of the roe is a beautiful, bright foxy red, which becomes thick and dark brown in winter. Two bands of white can develop across the throat in winter and a white rump patch, which appears in both sexes, is flared in alarm. Since bucks cast their antlers in November or December, an excellent way to distinguish a buck from a doe during winter is to look for the anal tush of the female — a long tuft of hair which is often incorrectly described as a tail. The male of the species has no anal tush. The buck should have grown its new set of antlers by about April or May when they should be free of velvet.

Tracks and Signs
The roe slot or track is usually 3-4cm wide and 4-5cm long. On soft ground their cleaves can grow fairly long, and narrow toward the front of the track. On harder ground, such as on Highland hill ground, their hooves tend to be more worn and stubbier. When walking normally their track is small and neat, with both cleaves closed together. However, on soft ground or when the animal is running, they splay out, giving a marked forked appearance. The imprint of their dew claws is visible on soft ground. The normal track left by a walking roe would be in groups of two, fairly close to the median line. Where they have been running you will find their tracks in groups of

four, splayed, with sign of the dew claws. The distance between each group can be more than 6ft.

Unless you have a crop like a thickly growing and succulent wall ivy where you would find obvious leaf stripping up to a height of about 4ft, roe do not leave many signs of their feeding other than their droppings. These are black, oval and 1.5-2cm long. However, during the rut they leave more signs of their presence. Small bushes or saplings, where a buck has taken out its rage, will often be scratched, scraped or completely flayed. Rings, or figures of eight, may be found where the buck has run in a circle around and around after the doe in part of the courtship ritual, although to see them mating is extremely rare.

Although they drop their antlers between November and December, it is unusual to find them since, like red deer, they normally eat their antlers for the vital nutrients they contain. Other signs you can look for are patches of bark which have obviously been rubbed or frayed on the bowl of large trees. These are favoured by bucks for removing velvet. Flattened areas where deer have been lying up would have a musky odour.

Muntjac *(Muntiacus reevesi)*
The muntjac are the smallest deer to be found in the country, establishing themselves from escapees of private collections, such as at Woburn, in the late nineteenth century. Today they are to be found locally throughout the Midlands and the south of England. They live singly or in pairs, and can exist quite happily without man ever being aware of their presence since they are secretive creatures and prefer to live in tall vegetation. Their small size of 40-45cm at the shoulder makes them difficult to spot, and a fleeing reddish-brown back is easily confused with a hare. Delightful little creatures, the bucks have small fangs and tiny antlers, and will give quite a sharp bark when frightened or aroused.

Muntjac tracks are only about 3cm long, pointed and narrow. Other tell-tale signs of muntjac can be regularly used tracks which through long vegetation give the appearance of tunnels. Their droppings are black pellets less than 1cm long, and are normally found in regularly used latrines. The muntjac's odour is distinctive, particularly on rubbing points used by bucks.

Although legally they may be shot it is a pointless exercise unless they are numerous within your area and need to be culled. That would be rare since they do little appreciable damage to arable crops so there is no justification as far as crop protection is concerned. They fall into the category of species that should be encouraged to enrich our environment.

Antlers

Deer antlers are often wrongly referred to as horns. Horn is a different material entirely, which grows permanently over a central core, whereas antlers drop off once a year, to be completely re-grown, and consist of bone made up of large quantities of calcium, magnesium, and phosphorus. The main function of antlers is to display rank and stature, and a much lesser purpose is as weapons of defence. Antlers become larger every year, until the animal's eventual decline through age, when they are referred to as 'going back'. This condition can also occur through ill health.

The antlers of young deer, of whatever species, are small and plain affairs in their first year, usually consisting of just a single branch. With each new growth of antler the spikes will become more elaborate, gaining thickness and weight until the animal is fully matured. At what age a head will start to 'go back' depends on the species itself, and on its feeding. A red deer with good feeding will, generally speaking, be fully grown at around seven years old, and will start to go back at fourteen. A roebuck will be fully grown at three or four, and will begin to decline by the time it reaches seven or eight years.

While growing the antlers are covered with a light skin and dense, velvety hair, known as velvet. At this time the antlers have an excellent supply of blood, both through the core and just under the velvet. It is the blood vessels under the velvet which result in the animal having good pearling, the bony ridges on either side of the arterial channels, which can be seen when the antlers are cleaned. In growth antlers are easily damaged, and if they are, will not only cause the animal considerable discomfort, but will be distorted or stunted for that year. When the antlers have reached their full growth for the year a ring of bone grows out from the bottom of each antler at the top of the pedicle, which is known as the coronet. This cuts off the blood supply and allows the antler to harden. It is at this point that the animal starts to clean the velvet from its antlers as it dries and falls away. During this exercise the deer becomes familiar with the shape and size of its new antlers.

Cast antlers are rarely found on the hill by walkers or stalkers. This is because they consist of such a rich quantity of minerals that they are eaten by the deer themselves. This is particularly the case in lime-free terrain, such as the Highlands and mountain areas of Scotland, where the deer will chew the antlers for their vital nutrients. On several occasions I have witnessed hinds chewing the antlers while they were still attached to a stag which was lying in the heather! Deer will sometimes even eat the dried bones of long dead rabbits and sheep for the same reason. On islands such as Rhum, off the west coast of Scotland, where the deer have ready access to seaweed, they are less likely to consume as much antler, and consequently proportionately more casts can be found in that sort of locale. Equally in heavily wooded areas such as the south of England, which is rich in lime, the animals' need to eat the antlers is diminished, resulting in fallow antlers being found more commonly.

The physiological aspect of the growth of antlers is really quite impressive. The antlers of red deer, for instance, can weigh, when fully grown, about 25 per cent of the stag's total skeletal weight, or 5 per cent of its total body weight, this growth being achieved over just three to four months. Since antlers are directly connected to the animal's hormonal balance and testosterone levels, it is interesting to note that damage to an animal's testes by inherent hormonal imbalance, physical damage caused by fighting, or barbed wire, may inhibit the growth of antlers, and create in red deer what is called a hummel. Hummels can also occur naturally, and therein lies a dilemma for the deer forest manager, since he has no immediate way of knowing whether the animal is sterile but capable of normal sexual urges, or whether it is entirely fertile. The problem is compounded since in some animals the minerals which would have naturally gone to create antlers go instead into the animal's bone structure. It is not unusual for a stag which may have grown to 240lb in its prime, to go to over 300lb, with the bulk of this development going into the fore-quarters, making it almost bull-like in appearance.

A hummel may have very pronounced sexual urges during the rut, taking and holding a large herd of hinds, and fighting off smaller competitors. It is in effect holding the hinds until they are past their own period of fertility, and if it is not fertile those hinds will be without calves in the following year, with obvious detrimental effect on the herd's regeneration. Yet hummels can be entirely fertile and able to serve many more hinds than the largest stags, but since there is no real way of knowing the state of a hummel's fertility they are normally shot to enable better quality antlered stags to have access to the hinds.

Red deer stag

7 STALKING

The art of stalking any animal is to get as close to it as you wish, unobserved by the animal, and remain so until you have achieved your purpose for being there, whether it is observation, photography, or shooting. To be a good, careful and productive stalker requires not only an intimate knowledge of the environment, but a real feeling for what you are doing, control over your body and muscles and a delicacy and stealth. Most of us can be taught the rudiments of stalking sufficiently to get us by, but the really great stalkers are those who have an inherent feel for it, which I do not believe can be learned.

Any Australian who has ever worked with aborigines out in the bush will recount tales of how they seem able to follow an invisible track across mixed terrain, including rocks, or even the smallest creature, and to a white man's eyes it appears to be magic. In North America the Indians were masters of the stalk, particularly tribes such as the Apaches, who developed their stalking instinct to a high degree because of their extremely inhospitable environment of sand, scrub and rock. No such luxuries for them as footprints in soft soil, broken twigs and so on. The famed Gurkhas and the British Special Air Service have the apparent magical ability to travel through all terrain, including thick jungle, silently and invisibly. Yet none of these amazing stalking abilities have any superhuman elements. It is purely the fine honing of all the senses, and a high degree of physical fitness and awareness of fitting themselves in and becoming part of the landscape, not invading it by charging along reeking of after-shave lotion, tobacco smoke or toothpaste. I have watched heavy African lionesses set up a perfect ambush on antelope, their great dun-coloured bodies disappearing into the short dry grass, to appear magically ever closer to the unsuspecting yet alert prey.

All these examples involve the simple rules of stalking. First,

129

remember that movement can be detected by even the most sluggish eyes. Second, by using available cover, blend into the landscape, never standing out openly. Third, be aware of colour. Whether the watching eye is human or animal, as it sweeps its surroundings it is looking for something out of place in the picture, whether it be a red cap, waving hand, or flashing reflection. Fourth, wind direction — approaching an animal from the wrong direction is pointless.

The stalker must be constantly aware of every aspect of his surroundings, noticing even the tiniest obstacles in advance, and then move around them.

At all times you must enter a woodland or begin a stalk from downwind. Now in some cases this may mean a considerable detour before you can put yourself in a starting position to enter your ground with the wind on your face. Yet it is absolutely vital. There is no point whatsoever in entering any area you intend to stalk with the wind blowing the dreaded man-scent in front of you, for it will warn all creatures of your impending appearance, and they will quietly slip away. But if that is not possible, for reasons of geography or trespass, it is possible to stalk with a sidewind, ie the wind blowing across the way. However, this increases the possibility of your scent putting up any animals downwind of you, makes your stalking more difficult, and places an even greater emphasis on the need for stealth.

Red Deer Stalking

There are many estates throughout Scotland that rent stalking by the day or week, for both red deer stags and hinds. Stag stalking is certainly the most expensive, and hind stalking, though gaining in popularity, is considerably cheaper.

It is generally thought that stags are more difficult to stalk, but this is not necessarily the case since their groups are small and consequently easier to get at. In the rut they tend to be partially preoccupied with other matters. Conversely, hinds live in much larger groups and that, if for no other reason, makes them more difficult. Not only are they more wary, but also it is obviously more difficult to approach 100 animals than half a dozen. Also, hinds are led by a herd leader — an old, experienced beast which lets little get past her.

When deer lie down they lie with the wind at their backs. This means that not only is the wind bringing any scent of potential danger, but their eyes are fixed in the downwind direction, while their ears continuously swivel like radar for an unusual sound.

If you decide to go stalking in Scotland for red deer stags, it is advisable to spend some time on your general fitness beforehand, since nothing spoils a trip to the mountains more than the inability to face the rough and difficult terrain. The normal practice on most estates is for the professional stalker to take you on the firing range prior to going out on the hill. While it is an obvious relief to the stalker if you can give a respectable grouping on the target, if you are off honesty is the best policy and no matter what excuse for missing you can dream up, you can guarantee that he has heard it before!

Once the stalker and you are satisfied with your competence you will set off to the area he intends to work that day. This can involve either a long walk before starting the steep climb, or a short Land-Rover journey, if there are access roads for the vehicle, before you set off up the hill. The stalker will then 'glass' the area, looking for a shootable beast. Do not expect to go after a stag with a rack of antlers like Christmas trees. Only the most inexperienced and irresponsible stalker would even contemplate shooting the best heads, and although there is always the lucky exception where an old stag past its best has a really super head, it is the stalk and the whole experience you are there for, and the relative size of the antlers should be of little importance.

When the stalker has spied a beast he will normally plan an approach to give you — the guest — the best opportunity to get into a good position from which you can take the shot. Most stalkers worth their salt will explain, if asked, why they have decided on a particular approach. Normal obstacles to be overcome are wind direction, the presence of other deer on the ground, geographical difficulties, or a combination of all of them. Remember that while stalkers tend to be as fit as mountain goats they are usually sharp enough to recognise the guest's physical condition. There is no point in struggling up a mountain puffing like a steam engine. Take your time and the stalker will match his pace accordingly.

Once the stalker has you in position he will give you time to get your breath back and compose yourself before inviting you to take the shot. Once again take your time. It is at this point that the less experienced make simple mistakes — concentrate on your breathing, try to relax, and take the shot. In the event of the beast being wounded, either shoot again or hand the rifle to the stalker, since your first duty is always to kill the stag quickly and cleanly. The stalker will then perform the gralloch, although it is my belief that an ability and knowledge of the task is necessary for everyone who practises the sport, so offer to assist, or better still ask if you can do it yourself — under his direction if necessary, and offer to help in dragging the beast to the nearest access track where it will be collected. The stalker may be more willing to impart his knowledge to someone prepared to share the less pleasant parts of the day.

One other important factor to keep in mind — stalkers are not well paid and it is the custom to tip them — about £10 per day is the norm, at the moment.

If on the other hand you are fortunate enough ever to get out on the hill on your own there are a few simple guidelines for you to follow. First of all familiarise yourself with the contours of the hill, and your march lines. There are few fences in the Highlands, and the divisional line between one estate and another can be as vague as the bottom of a valley, or the course of a stream. Having satisfied yourself as to both the lie of the land and where deer are likely to be found, approach the area with a maximum of caution, keeping wind direction as your first priority. It is essential, as I have already said, that you always work into the wind, otherwise you may discover that deer a mile away from you have already been given prior warning of your presence and have moved, or are alert and fidgety.

One of the most important pieces of any stalker's equipment is his binoculars. It is essential that before you move on to any ground you carefully and systematically glass the hill, checking for the presence of animals, including sheep, and only when you are quite certain of the location of any beasts and have worked out a route which will take you past them out of both scent and sight, should you move. Do not try to short circuit this exercise. It is better to accept that you may have to make a long detour, giving animals a wide berth, than to think that you can sneak past them. You won't.

Once you have glassed the hill and spied a stag which may be suitable for shooting, work out your approach route, again making careful note of the presence of any other animals on the ground, and only when you are entirely satisfied that you have planned the best method of approach should you proceed with your stalk. Keep yourself as low as necessary, using the contours of the ground, and get yourself as close as you can. Ninety-nine times out of a hundred this invariably means that you are about 300 yards from the beast when you must lie down and start to crawl. Irrespective of whether it is water or snow, lie down. Do not try to keep yourself dry by raising yourself that extra inch. It won't work. Lie flat, and work yourself forward as slowly as you can. Do not be tempted, when you get to a range where you think you can hit the deer, to attempt a shot. It is much better to work on the maxim that you cannot be too close when you take your shot. The golden rule at this crucial stage in any stalk is taking your time. Move forward slowly, trying not to get out of breath, and when you are well within range and are quite certain you cannot get closer, lie still and prepare yourself. Hurried and snatched shots are seldom successful. Be honest with yourself. If you are an excellent shot and your rifle is pin-point accurate and precisely zeroed, and if the beast is either lying down with his head up, or standing side on to you, then the best place to aim is at the base of the neck. Struck in this position, with its spinal cord broken, the beast dies instantly, with virtually no wastage of meat. If on the other hand there is the faintest doubt whatsoever in your own abilities, or your rifle's performance, shoot the beast behind the shoulder, through the heart and lungs. The fatal area at this point is much larger, giving you room for error of several inches in all directions.

It is at this juncture that experience really counts. An animal shot through the heart and lungs can drop instantly or, more alarming to the uninitiated, gallop off for a hundred yards before going down. So unless you are experienced and know what you are doing and can read the animal's reactions to being struck, if it starts to run shoot it again. As soon as the animal is down go up to it and, unless the head is to be mounted by a taxidermist, cut its throat high up under its chin, to prevent meat wastage. If the head is to be mounted, I do not advise the cutting of the throat at all.

If you approach a deer you have shot which is obviously not dead but struggling to get up, it is advisable to shoot the animal again. However, in the normal situation when an animal is not dead, it will lie gasping, perhaps lifting its head or kicking. Then it is unnecessary to shoot the animal again. However you must be positive. Go quickly to the animal, approach it from behind, stretch its head back and cut its throat under the chin, paying particular attention to the area on the left and right of its windpipe, and cut it until your knife touches the spinal cord. Then you will be in no doubt that the job has been done correctly. If the animal has antlers and there is a danger of it catching you, then once again approach it from behind, grasp the antlers firmly and holding the head back, cut its throat.

133

It is important that you start to gralloch the beast immediately. Roll it onto its back and neatly cut off its testicles and scrotum. It has become almost tradition in the Highlands for you to say as you perform this task 'You'll no be neddin' them nae maer!' You will find that after you have made this incision the stomach wall should still be intact. Carefully slit the skin, taking care not to open the stomach, and run your knife up to just below the sternum (breast bone). Then using your hands, flay the skin back around the stomach. It is actually possible at this stage, after a little practice, with your hands alone to loosen the skin all around the body by pushing your hands through the warm dermis — that is the tissues which hold the skin to the body. The reason for doing that at this early stage is that later in the larder the skin will come off so much easier when you butcher the beast. However, it is not absolutely necessary to do this, and it is certainly easier with the smaller species of deer. Also, it is only really desirable if the carcass is intended for your own larder. If it is to be sent off to the game dealer you need not bother as the skinning of the beast will be done by the game dealer himself.

When you have flayed the skin back from the stomach, carefully make a small incision in the stomach wall. Then insert the index and middle finger of your left hand (if you are right handed) and pull the muscle tissue upwards, placing your knife carefully between your two fingers, and slit the stomach wall all the way up. This manoeuvre sounds complicated, but is in fact quite simple. All you are trying to prevent is slitting open the intestines, which is not only messy but also the bile can taint the meat.

Once you have slit the animal — and again, this method, which is to my mind the best, is specifically for the man who intends the carcass to be for himself — put your left hand inside the rib cage and you will feel the diaphragm, the muscle wall which separates the heart and lungs from the intestines. Cut through the diaphragm carefully and keeping the back of your hand against the sternum, push your right hand up to the throat, where you will feel the windpipe which you will grasp in your left hand. Then carefully reach in with your right hand, cut the windpipe and withdraw the knife.

You will now find that if you grasp the windpipe in both hands and pull you can strip the beast completely, all the way down to the pelvis. Pull the intestines out and dump them between the hind legs. Clean the anal tract out through the pelvis, taking care not to puncture the bladder, which you must hold by the neck, and cut free.

Now carefully retrieve the two kidneys, the heart, and the liver. On the inside surface of the liver you will find a small, greenish-purple, tube-like vessel. This is the gall bladder and must not be burst. Cut it carefully out of the liver and discard it. Putting heart, kidneys and liver aside, slash the intestines and stomach all over. This prevents them gassing up and not only assists predators in eating them but also helps them to rot away quickly.

The beast is now ready for transport to the larder where it can be headed and hocked. Remember, if the head is to be mounted by a taxidermist cut off plenty of neck skin with it.

Woodland Stalking

Woodland stalking is entirely different from hill stalking. In the woods you have much more vegetation to stand on or brush against, causing disturbance and announcing your presence. There is a greater likelihood of small birds, jays and other creatures observing your passage and calling in alarm. On the other hand, in woodland you have the advantage of the trees giving you cover, and reducing the area in which you may be seen.

If you have taken all the necessary precautions in the selection of your equipment you should arrive at the woodland clothed in soft, non-scratch, non-rustling material of muted colours, comfortable enough for you to be able to bend or crouch without the materials restricting you, with no rattling or squeaking leathers or accessories. The great Howard Hill, guru of all bowhunters, once, as a young man, asked a Seminole Indian what was the secret of successful stalking, and was given the answer, 'Walk little, look much'. That advice should be the rule that every stalker follows.

Each ride, glade, or clearing you are about to enter must be scanned carefully with your glasses for that tell-tale ear-tip sticking out of the long grass, announcing the presence of a couched roe doe, or a group of rabbits which are going to whizz off on your arrival. If the obstacle cannot be circumvented you must make a decision either to sit and wait until it has moved on by itself, or to gently show yourself. In the case of rabbits it is not so serious, but a startled roe doe will bound off barking in alarm to alert every buck in the vicinity. At all times move with great caution, putting each foot down carefully on the ball, rolling it down onto your heel, and then transferring your weight to that leg. It takes a little getting used to, but it certainly pays results.

When you see a buck in the distance take your glasses and study it carefully. Is it a shootable animal? Once you have decided that it is plan your approach route carefully. Get yourself as low as possible. It is not sufficient to crouch and think you are safe, always err on the side of caution, make yourself even smaller. Do not try to avoid the muddy puddle. If it means you must crawl through it on your belly, get wet. Do not risk the beast seeing you by merely crouching to get over the puddle. By this means you will succeed in most cases. Take your time, keep your face down, and try to follow this simple routine — if you can cautiously take a peep at the animal when its head is down move forward and lie still, then take another cautious look. Deer are extremely nervous and alert animals, and will never stand grazing like a cow, but instead take little nibbles and bring their heads up again. Their eyesight is not particularly good, but movement they can detect instantly. So, if you find a deer looking straight at you, you must freeze. Move not a muscle until the beast is satisfied that you are not a threat. When it resumes eating, cautiously move forward again.

When you have finally got within range, lie still, make sure your breathing is normal, then ask yourself, can you get closer? The aim of every stalker should be always to get as close to the animal as possible and never to be content with just getting into range. I can group bullets from my .243 or .270 rifles neat as a wink at 150 yards. Yet the majority of roebucks I have ever shot have been within 50-80 yards away, and some even closer than that. After all, the shooting of the animal is really an anti-climax. The real thrill should be in the stalk. Shooting the animal ends the stalk and converts it into meat for your deep freeze.

Once you have made the decision to shoot take your time, never rush. Unless you have absolute confidence in your own marksmanship and your rifle is perfectly zeroed, do not shoot the animal in the base of the neck. Better to go just behind the shoulder and kill than take even the slightest risk of wounding or missing.

After you have shot the animal, go to it and make sure it is dead, then start your gralloch. If you intend to continue stalking that morning, when you have completed the gralloch lay the animal on its stomach with the hole propped open to allow it to drain, for collection later in the day.

Flying mallard

8 PHOTOGRAPHY

Most field sportsmen eventually reach the point where they begin to regret their inability to capture some of the scenes witnessed on film and their not having a camera, even if just to record an event, a bag, or a specific species. Most people who are not keen on photography regard it as a mystical semi-scientific minefield that is not for them. Yet nothing could be further from the truth, and I believe that one of the more important items in every outdoorman's equipment is a camera. A camera is simply a black box into which you load film, and by momentarily letting light in through the shutter the image you are viewing is frozen on the film.

There are two classes of camera. The cheaper models have fixed-focus lenses built into the body. Though these can serve very well for many years they have limited use and with more experience you will need greater versatility. Having discovered that photography is really very simple, for more adventurous work you will require something more versatile, probably necessitating a completely new camera and substantial cash outlay. This could have been avoided if the correct camera body had been bought in the first place.

Amateur photographers the world over have reached such a high level of proficiency and excellence that they demand a high standard of equipment, and it is a fact that there are not sufficient professional photographers in the world to keep one camera manufacturer in business. All camera manufacturers make their top range of cameras to professional standards while still selling them to the amateur. It is only when you come down the scale of price and sophistication that manufacturers cater for the 'impulse' photographer — the person who knows nothing about it, and does not wish to learn, and whose requirements are completely satisfied with a modern version of the old box Brownie. Yet these cheap cameras mostly use the same film as their far more desirable stablemates, and that film costs exactly the same to process. So if you buy one of the better cameras, after the initial outlay the running costs will be exactly the same as if you had bought a cheapie, but the results will be far superior.

A few simple rules are all that the field sportsman needs to guide him. First purchase the most expensive camera body that you can afford, with one or two good lenses. As you become experienced and familiar with your equipment, you will probably develop particular interests and ideas.

The best advice for anyone about to buy a camera and take up photography is to understand that the more sophisticated camera bodies are normally designed as the centre of a system — the body is the heart of the equipment, and has a wide range of interesting accessories which may be attached to it, giving the photographer an ever-increasing range of possibilities. Let me give you an example. A friend of mine purchased a camera for a little over £100, only to discover very quickly that he was limited in the type of use he could put it to, being restricted to variations of 'happy snaps'. My equipment has at its heart a very good camera body which cost a little over twice the price of my friend's camera. I can attach a motor drive to it, allowing me to take up to five frames in a second, a very desirable piece of equipment when photographing moving subjects. I can fit a wide range of lenses to cater for every requirement. I can even attach a small receiver to my camera, prop the camera up on a rock or in a tree beside a peregrine's eyrie for example, then, sitting comfortably a quarter of a mile away, at the press of a button on my transmitter can take photographs. One must remember, however, that when photographing certain rarer species of wildlife licences are necessary, and you should contact the Nature Conservancy Council for details.

It is vital that the beginner buys the best body he can afford. Then he has the security of knowing that irrespective of how keen he may become as his interest develops, his equipment will always be capable of doing the job.

With a little practice you can quickly get used to handling a camera, and will soon be making the best of all weathers. Rain, for instance, can give the most delightful atmosphere in photographs, and as long as you remember carefully to dry your camera as soon as possible on returning home, it should present you with few problems. Obviously you must take care. If shooting in the rain I usually keep my automatic camera inside a plastic bag, with a little hole cut out for the lens. I have become quite used to carrying my camera under my jacket. In mist or snow, or when the sun comes out after a damp morning you will soon learn to recognise the different textures of light, and the sort of effect they will give you, capturing the mood of the moment and freezing it forever.

Hide Photography

If you are going to take photographs from a hide which you also intend shooting from, such as of geese coming in to decoys, then it is best to have your camera ready, mounted on a tripod, with a hole cut in your hide net to accommodate the lens. You must decide in advance when you are actually going to take the photographs since it is not advisable to try to shoot and then grab your camera.

The simple technique most people employ is to shoot a few birds, then put the gun down and leave the next few geese for the camera alone. If you are taking photographs of wildlife from a hide you do not intend to shoot from you are obviously going to get better results. Then you build the hide and leave it so that the game becomes used to this new feature in the landscape. This is particularly important where you intend to photograph nesting birds or animal dens, or indeed anywhere where the game may be extra alert. Once the hide is established you would enter it before dawn, and when you are familiar with it you should be able to do so with great secrecy.

Camera Consciousness

Start making yourself as camera-conscious as possible, otherwise you will discover too late that your moment has gone before you remember your camera is slung around your neck. When I first started carrying a camera, two events took place which I will never forget. While stalking roebucks one early morning I spied a good buck in a grassy field. As I started to stalk it I realised there was a doe lying in reeds close to where the buck stood. I made my way carefully until I was within 80 yards of the buck, and lay flat on my stomach, preparing to take the shot. The buck aroused the doe and started running around in a tight circle with its nose under the doe's tail, making snuffling sounds. Then, abruptly, the doe stopped and the buck mounted her. I couldn't believe my luck in witnessing such a rare sight, and lay fascinated, only partly aware of the uncomfortable rock pressing into my ribs. But since the cover was so short I decided to remain in this uncomfortable position rather than disturb the beasts.

Over the next two hours I lay in the same position, fascinated as the roe went through their mating procedure. First the doe would lie down and the buck would nibble nervously at the grass. Then the buck would lie down for a short time before bouncing back to his feet, prodding the doe out of her couch. I was quite confident I could shoot the buck at my leisure, but was more fascinated by witnessing such a rare sight, one which few men are ever privileged to see. Seven times in all the buck roused the doe off her couch and went through the same procedure before serving the doe. Suddenly a vehicle passed on a distant road, and in a flash the two animals disappeared. Thankfully I stood up off the accursed rock which had been grinding into me, to discover it was my camera! I had actually witnessed such a rare event, photographs of which would have been extremely valuable, and hadn't even thought of using my camera. Indeed, I had completely forgotten I was carrying one.

On another occasion during that summer, again while roe stalking early in the morning, I witnessed two eagles playing in the thermals, plummeting like stones from a high altitude, and just before striking the trees below opening their wings to rise effortlessly up on the rising currents. Continuously while the birds were rising they were being attacked and struck by three peregrine falcons, two of which had a nest close by. Through my binoculars I could actually see feathers being knocked out of the eagles as the peregrines repeatedly dive-bombed them. I made my way high up to the perimeter of the bowl and sat down in the long heather to watch this incredible sight. One eagle passed me close in its stoop, and I remember clearly its gold mantle glinting in the sun as it turned its face and looked straight at me as it plummeted down. I could have screamed when later I discovered my camera neatly packed in my small back pack.

These are two extreme examples of learning the hard way. Had I been more camera-conscious, both sets of photographs, apart from being most memorable, would also have been extremely marketable. Since these events I have become accustomed to carrying my camera where it is easily got at, no longer preciously wrapped in its leather, 'never-ready' carrying case. It is a piece of equipment always at hand and ready for action.

While it is a good idea to take your camera with you whenever you go shooting, it is a compromise, and for most people shooting will

A red deer hind lying down; although she is watching in one direction, her ears swivel round in response to a sound from behind

A well-hidden red deer, camouflaged in long grass

A superb red deer stag in velvet

take precedence over photography. If you are going to take photographs, leave your gun at home. The temptation to use your gun is always there and it is preferable to concentrate on your camera. You will soon discover that your camera will give you a completely new aspect to wildlife study and heighten your appreciation of the outdoors.

Composition of Photographs

Whenever you are taking a posed photograph — of your shooting companion or your dog with a bird — the key word is imagination. It is not sufficient to have your chum stand rigid, clutching his brace of birds, grinning self-consciously at the camera. Use local features and introduce as much colour and tone as possible to create a mood. A brace of pheasants with their feathers carefully smoothed down, lying on golden stubble, with your dog, broken (open at breech) gun, and two empty cartridges, are more interesting than two pheasants carelessly flung down, feathers ruffled, wings akimbo on some trampled grass with your dog standing over them looking as though it is about to eat them.

Once I took a highly emotive photograph of a roe stalker, partly obscured by a screen of leaves in his camouflage suit. He merged completely with the background and at first glance was almost invisible. Yet the mood of the photograph emphasised the stealth and desirability of a roe stalker to merge with his surroundings. When I took the photograph the fellow kept suggesting he stand in front of the bush, so that he could be seen better. On another occasion on top of a mountain I was photographing a stag I had just shot. With my back to the valley below, the photograph would have been of a stag lying on heather, no different from any other. Yet by walking to the other side of the stag so that it was in the foreground, with the great valley below and the distant mountains rising beyond, the photograph illustrated perfectly the loneliness and solitude of the environment. Simply a case of using imagination.

Equipment

The Body Recommended manufacturers are: Nikon, Canon, Minolta. Choose a camera from near the top of these manufacturers' ranges. It is desirable, wherever possible, to have an all-black body.

Lenses   Most cameras come with a 50mm lens fitted as standard. This is the lens which has the nearest field of vision to the human eye and is an excellent all-rounder.

Telephoto Lenses   These are long-focusing, fixed-magnification lenses. Their problem is that the field of vision becomes narrower in proportion to the magnification, and they are really unsuitable for general wildlife work unless being used for a specialist purpose such as photographing a distant nest such as a falcon's on a cliff. They are not nearly as versatile as zoom lenses.

Zoom Lenses   These give the photographer by far the greatest variation and versatility being, if they are of good quality, several lenses in one, enabling the photographer, with a twist of his wrist, to focus on a distant object with high magnification, or 'pull back' to give a more general view of the scene. For example, imagine you are standing at the end of a line during a pheasant shoot. On your camera is a 70-150mm zoom lens. You can, without moving, photograph one of the guns in close up as he shoots, one of the dogs retrieving a bird, or the whole general picture of the field. I would not recommend greater magnification than 200mm as beyond this it becomes difficult to hold the camera still without a tripod, since the greater the magnification the more hand tremor is emphasised.

Wide-angle Lenses   These give an exceptionally wide field of view and can be fun to use. However, the problem is that the wider the field of view, the smaller each part of the image becomes. Another problem with wide-angle lenses is that if photographing a close subject you tend to get distortion, or stretching of the image, which is obviously undesirable.

Filters   There are a wide variety of filters on the market, offering the photographer everything from starbursts to pre-shaped or heavily coloured prints. The use of most filters is the personal choice of the photographer, and only two need be considered immediately.

Every lens should be fitted with a skylight filter. These are cheap and filter out unwanted rays. Equally important, they also protect the surface of your expensive lens from scratches, dirt, general knocks, and so on. The skylight filter should be left attached to your lens all the time, only being removed when you wish to fit another filter which is not compatible with the skylight. Polarising filters are particularly useful to cut out light reflection from snow or water. As you become more experienced you will start to experiment with the results to be gained from using other filters. For example, when photographing snow you would use a blue filter since snow has a bluish white tinge and the blue filter gives more of a cold, snow-like effect to winter photographs.

Motor Drive This is a battery-operated motorised attachment which fits onto the bottom of your camera enabling you to take photographs in rapid succession. As you press the shutter release it winds your film forward, exposing between one and five frames in a second depending on which gear you have selected.

Power Winders These are cheaper, fixed-speed automatic winders to roll the film forward. The power-winder and motor drive allow the photographer to concentrate on focusing and keeping his eye to the viewer without having to break off to thumb the film forward. But the real advantage, of course, is when photographing moving objects.

Tripod This should be strong with firmly telescoped legs. Avoid the smaller pocket versions; they do not last and are prone to falling over at the slightest knock.

Film For black-and-white two speeds are recommended as a basis for the outdoor photographer. When photographing fast-moving objects, or in bad light, fast film of 400 asa should be used, though the quality of the photograph will be sacrificed through grainy prints. For general black-and-white work 125 asa is ideal.

With colour, once again for bad light or high speed objects faster film is advisable — 400 asa, though by far the best colour reproduction is from 64 asa. Kodachrome is excellent but has the disadvantage of having to be sent off to Kodak for processing. I have found Ektachrome to be equally suitable and has the advantage that it can be processed by any competent colour processor.

Do not use cheap film or processing. In the majority of cases much quality is sacrificed, and that little extra charge is well justified in the results obtained by good-quality film processed by professionals. Most important is that if you ever take photographs which you think are of a particularly important or valuable nature, do not send them off to be processed, but take them and ask for them to be specifically done by hand. Can you imagine, for instance, if you had film of the Loch Ness Monster and due to the over use of chemicals they were obliterated or damaged. It would not matter how much you jumped up and down in rage; they would be lost forever.

Author's Recommended Kit   Camera, motor drive, 50mm lens, 35-80mm zoom lens, 70-150mm zoom lens, cleaning tissues, tripod. All chrome or shiny surfaces should be masked with black sticky tape to cut out reflection.

Roe buck

9 TAXIDERMY

A growing number of sportsmen have, in the last few years, revived the practice of having good specimens of birds and animals they have shot mounted by taxidermists for display in their homes, offices etc, in contrast to the time when I first started studying taxidermy, when most specimens mounted were for museums, with the odd one or two for private collection. Sadly, however, with the growing popularity of having birds and animals mounted, a number of 'stuffers' have appeared. They are amateurs who are cashing in on the fact that taxidermists are relatively few. The 'stuffer' is neither trained in the art nor the secrets of the profession, though he manages to charge professional prices for non-professional work. However, with a little patience and skill almost everyone who does not have ten thumbs can mount his own trophies, saving a lot of cash and getting a great deal of satisfaction from the finished work.

The tools you will need are:

1. a scalpel
2. rose secateurs or bone cutters
3. arsenic or alum and borax
4. wood wool
5. linen thread
6. small knife
7. killed plaster (plaster soaked in petrol and dried)
8. gun tow
9. thin string
10. thin tweezers
11. phenol
12. calipers

For your first attempt at skinning choose something simple, such as a starling. Lay the bird on its back with its head away from you, and part the feathers down the track on its stomach. (With sea birds and ducks skin from under their wing.) With the scalpel make a small incision down the belly from just below the breastbone to about half

147

an inch above the vent, being careful not to cut open the stomach. Work the skin carefully off the body down to the root of the tail, and cut through towards the back. Next, skin down to the bird's hip joint and cut the leg from the body. If the skin is getting slippery sprinkle a little killed plaster over it to keep it dry. Now work the skin off up the wing joint and cut through. The skin will now come off the neck fairly easily to where it joins the skull. Cut the neck through and lay the body aside. Carefully skin the skull down to the beak, making sure not to cut the skin on the face. With the scalpel clean all flesh from the skull and scoop out the eyes and brain. Peel away any flesh and fat from the skin, particularly on the feather tracts, making sure you do not cut any feather sheaths. Take care that you do not stretch the skin at this stage, otherwise it will cause problems later. Clean all flesh from both leg and wing bones.

When the skin is free of all fat, wash it in petrol, preferably high-octane aircraft fuel, to remove any oil or grease. When absolutely clean squeeze all the petrol out of the skin and dust it heavily in the killed plaster until it is completely dry. Then lightly beat out the skin until all the plaster has been removed. Rub arsenic or 50/50 alum and borax into the skin and lay it aside, covered with a damp cloth.

With wood wool or sisal make a body using the carcass as a model. Bind the body firmly with linen thread. Sharpen a wire at both ends and push it into the body where the neck is to be. Crimp the wire into the body to make it firm. Now bind the wire with gun tow.

When you are satisfied with the new body pass the neck through the neck skin and pierce the skull with the wire. Try closing the bird's skin around the body. similar to fitting a jacket. If the edges of the incision you made fit neatly, proceed. If not, alter the body until you are satisfied it does.

Lightly bind gun tow around the upper leg bone to simulate the thigh, then push sharpened wire through the insole of each foot, leaving 2in protruding from each leg. Keep the wire running up the back of the leg bone and crimp the wire into the body. Lay the bird on its back and sew the skin with waxed thread.

Now with your hands shape the bird into the desired position and place the foot wires through two holes on a piece of wood. The bird has now been mounted. All you have left to do is fix the wings into your chosen position using 2in straight pins to fix them through the joints. Model the head by placing small wisps of cotton wool through the eye holes, with thin tweezers. When you are satisfied with the position of the bird leave it to dry for two or three days. Then dampen the skin around the eyes and place glass eyes in the sockets, fixing them with a little glue paste. Remember that practice makes perfect. With patience you will quickly move to more difficult bird subjects.

Small Mammals

Before you embark on mounting your first mammal it is advisable to have mounted a number of birds to your absolute satisfaction so that you get the feel of what you are doing, since mammals, large or small, are much more difficult.

Probably the best small mammal to start on is a rabbit — not a large, pregnant doe, as the stretched skin and swollen mammaries would present problems. Preferably choose a three-quarter grown male. Carefully measure the circumference of the animal's chest and stomach; the length from the occiput (back of skull) down its spine to the root of its tail; and from the tip of its nose down its spine to the root of its tail. Then measure from the top of its pelvis, following the contours of the back of its leg to its foot; and finally from the spine between its shoulder blades, following the contours of its leg to its foot. These measurements are best carried out with a tape measure. The reason for taking these measurements is to avoid over-stuffing later on, when the skin is pliable, and the temptation is to stuff it until it is taut.

Lay the rabbit on its back, and with the scalpel cut a small line from the sternum (breastbone) down the centre of the stomach to one side of the testes, taking care not to penetrate the stomach, thus avoiding mess, and keeping the skin clean. Carefully peel back the skin, using your fingers and scalpel, around either side until the skin is free around its back. Then carefully work it down towards the root of its tail. When the skin is free behind the hip joints, cut carefully through the flesh and joints so that the hind legs are separated from the body, which may now be drawn through the incision in the stomach skin. Holding the pelvis in your hand, peel the skin up toward the head until you reach the shoulder blades where the forelegs join the body, and free them by cutting neatly between the shoulder blades and the body. This leaves the forelegs, like the hind legs, attached to the skin for the time being, as you continue peeling the skin up the neck toward the skull.

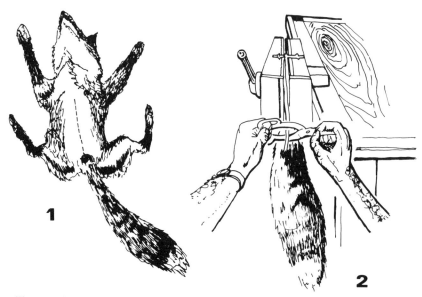

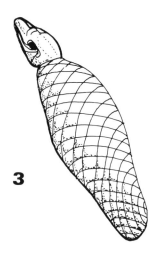

To start skinning, cut along the broken line

Remove the tail bone

The skull fitted to the body model

Making the limb muscles

5

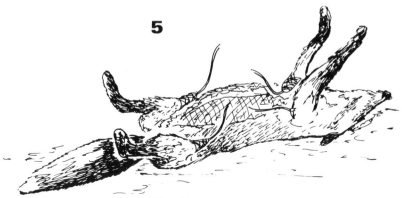

The skin ready to receive the body and skull

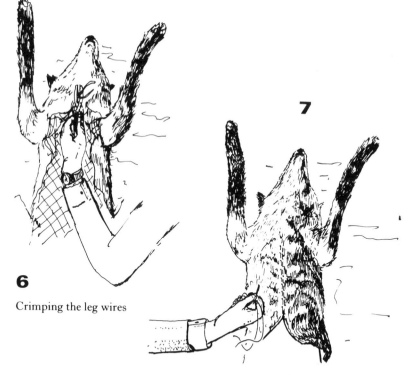

6

Crimping the leg wires

7

Sew up the skin

By this time the skin should be inside out across the table to your left, attached at one end to the head of the skinned, limbless body to the right. Carefully skin the skull until you reach the base of the ears, then cut through as close as possible to the skull, and carefully skin forward until you reach the corner of the eyes, which you cut carefully through, taking care not to cut through the eyelids.

Skin on until you reach the corner of the mouth (gape). Work your way along the inside of the mouth, where the inside lips are attached to the gums at the roots of the teeth. By this time the skin should be attached to the skull only by the nose. Free it by cutting through the cartilage around the nostrils as near the skull as possible, taking care not to damage the outside skin. Remove the skull from the body and draw a sketch of the skinned skull, adding measurements of its breadth, and notes on the muscle formation. Then flesh it and boil it, removing all traces of flesh.

If you examine the nose, which you have now turned inside out, you will see that the cartilage shapes it. It is important that you remove as much of the cartilage as possible, to prevent later shrinkage and distortion. Equally important is to finely split the eyelids and separate the cartilage from the inside surface of the ear from the back (fur side). Having completed this, which may take several specimens to perfect, skin the four legs down to the feet. Then carefully remove all the flesh from the limb bones, leaving only the cartilage around the joints to keep them together. Before you do this it is a good idea to lay both front and hind legs on paper and draw a silhouette of them as a later guide to shape, taking note of the thickness of the limbs.

For animals with long tails it is necessary to remove the tailbone. This is best done by covering the root of the tail with sawdust to prevent it slipping, and placing it in a vice. Then with the inside surface of the handle of a pair of scissors (see diagram) exerting firm pressure the tailbone will slide neatly out of the tail. It is important when pickling tails in the phenol bath that you poke a wire or thin stick down the tail, making sure the phenol penetrates it. With small animals and very small tails it is not necessary to skin them as there is insufficient flesh to prevent the phenol penetrating right through.

Proceed to remove all fat or tissue which may be adhering to the skin. When you are entirely satisfied that the skin is clean, turn it back the correct way with the fur side out and carefully check it for blood, urine or excrement. If you find any sponge it off with a little lukewarm salt-water solution. When this is done, place it in a phenol bath — a mixture of one part phenol (carbolic acid) to eight parts water, and leave the skin to soak for one week. Most chemists will supply phenol, though you may have to order it, so it is a good idea to purchase it in advance. If phenol is completely unobtainable it is permissible to use a mixture of 50 per cent alum and borax which you rub into the flesh side of the skin. Although this is not as suitable as phenol it will preserve the skin, which must be kept damp and pliable — but not wet — by being wrapped in a disinfected cloth.

Straighten and sharpen both ends of five pieces of wire, about ⅛in thick, preferably galvanised, of a length equal to that of the rabbit. When your skin has been pickled in phenol for a few days take it out and let it drain until all excess moisture has gone. (Some people prefer to wear surgical or household gloves at this stage to protect their skin from the phenol.)

Lay the rabbit out on its back. Carefully thread one end of the wire through the insole of a foot and thread it up the back of the limb bone, bending the wire to match the bone, leaving 3in of wire protruding from the paw. Carefully bind gun tow around the limb bone and wire. Make small muscles, modelling as near as you can an exact replica of the original limb, and tie them on with thread, using your silhouette as a guide. Repeat this with all the limbs. Now lay the specimen on its back, with all four leg wires protruding from the stomach incision.

Take the skull and use gun tow to model the small muscles you have removed and bind them on with thread. When you are satisfied with your work place it inside the head skin, lining up the eye holes on the skin with the eye sockets on the skull. Then take the fifth wire and pass it through the stomach incision, up inside the neck through the back of the skull and drill it through the forehead, leaving 2in protruding from the head. With animals which have long tails, such as a fox, which have had the tailbone removed, a wire one and a half times the length of the tail, two-thirds of it wrapped in gun tow to simulate the fullness of the actual tail, should be slipped down into the tail and the sharpened end passed into the body.

Now, referring to your measurements of body length, carefully twist all the wires together at the stomach incision with pliers and cut off the ends. Then proceed to soft-stuff the body with fine wood wool, taking care to pack behind the wires so that it is neither too empty nor oversttuffed, and stitch the incision neatly together. This stage requires care and practice. Then bend and model the specimen

153

into the position you wish, placing the foot wires through four drilled holes in a plank of wood. Your rabbit will now be standing roughly in the position of your choice. This requires both experience and artistry, but don't lose heart, with a little practice it will come to you.

For larger animals such as cats and small dogs it is necessary to make a solid body wrapped around the wire from the skull (see diagram) and crimp all wires from limbs and tail into the body. Both processes are the same, the only difference being in whether you soft stuff or have a bound body. It would be inappropriate to go into a long explanation here on other more advanced and professional methods employed when mounting larger specimens for museums. They involve space, equipment and considerable expense which very few sportsmen would be interested in.

Using tweezers, or a small piece of wire, put wisps of cotton wool in the mouth and eye sockets until you are satisfied with the head. Take two thin, sharpened wires and push them through the tips of the ears into the skull. Lightly place a small amount of modelling clay inside the ears for support. Then shape the eye surrounds, nose, and lips by placing small balls of clay inside the skin, and model them into a life-like position. When you are entirely satisfied with this cut the wires protruding from the tips of both ears and the forehead, and leave the animal to dry slowly at room temperature for at least a week.

You will then see from any distortion of the head if you have modelled each side evenly. Unevenness can be remedied by damping down the area and re-doing it. When the head is dry and modelled to your satisfaction, dampen the area around the eye with cotton wool soaked in water and any antiseptic liquid (Dettol, Savlon etc), and put the glass eyes in the sockets. Model the surrounds with clay and allow them to dry. Then carefully paint eyelids, nostrils, and lips to their former colour.

This method of mounting is known as 'soft-stuffing' and may take several specimens before you get the hang of it, but remember practice makes perfect.

Mounting Heads

The most difficult method to describe is the mounting of a head, whether it is a fox mask or a deer's head. However, the following description should be comparatively easy to follow, particularly when you have a specimen to refer to.

The first step in mounting any head is to take accurate measure-

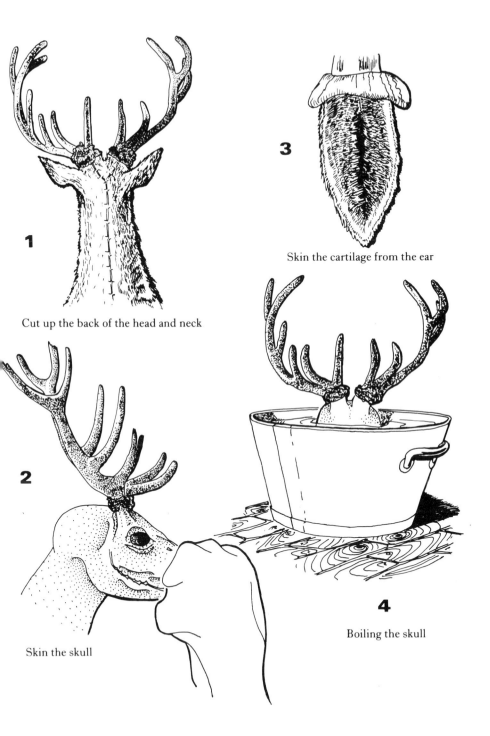

1

Cut up the back of the head and neck

2

Skin the skull

3

Skin the cartilage from the ear

4

Boiling the skull

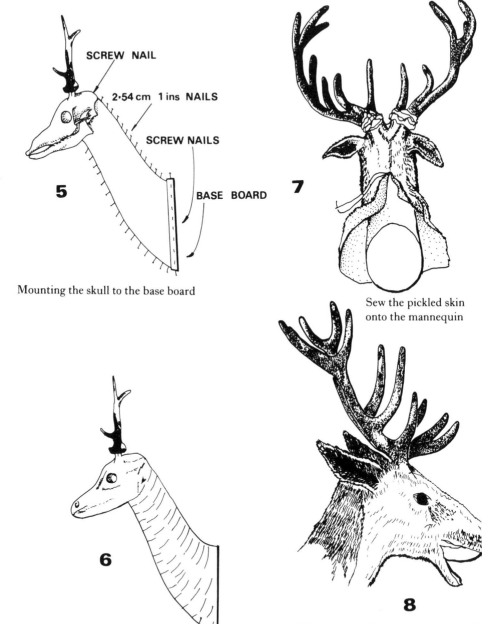

SCREW NAIL

2·54 cm 1 ins NAILS

SCREW NAILS

BASE BOARD

5

Mounting the skull to the base board

7

Sew the pickled skin onto the mannequin

6

The made-up neck and skull, with all the muscles in place

8

The mounted head, ready for modelll

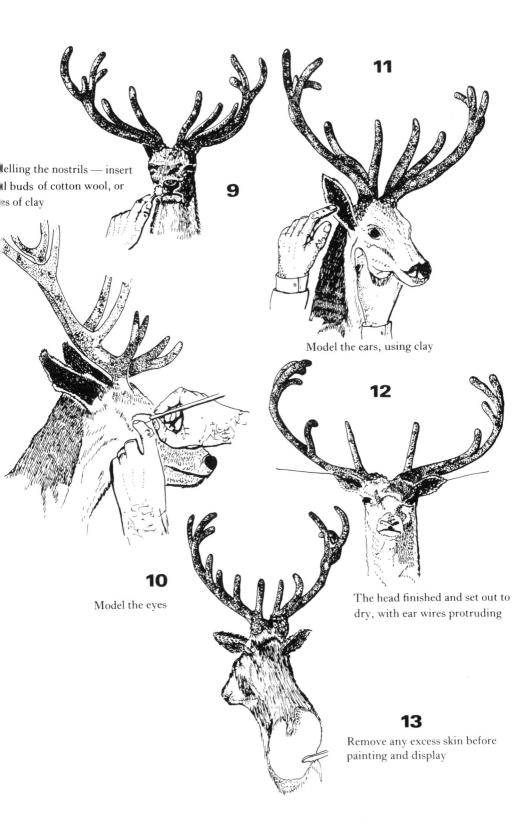

delling the nostrils — insert
buds of cotton wool, or
s of clay

9

11

Model the ears, using clay

10

Model the eyes

12

The head finished and set out to
dry, with ear wires protruding

13

Remove any excess skin before
painting and display

ments: tip of nose to corner of eye; corner of eye to base of ear; corner of eye to corner of eye; breadth of nose; depth of nose to chin; breadth of skull; circumference of neck, and so on. Use calipers and a piece of string knotted at one end. I have always preferred to use string since it enables you to follow the contours of the specimen more easily. There is virtually no limit to the number of measurements you can take, and I am of the opinion that there is no such thing as too many measurements.

I will not describe how to mount a head with its mouth open — a stag roaring or fox snarling, since to do this properly requires a great deal of experience in casting the tongue and modelling lips, gums and mouth. Nothing looks worse than a mounted head with open mouth, unless it is absolutely 'spot-on', and this is not easily learned without professional tuition. However, mounting a head with mouth closed is not quite so difficult.

Probably the most common mistake made by sportsmen is not retaining sufficient neck skin for the taxidermist to work with. When skinning a head it is essential to start the cut as far back as the shoulders around the base of the neck, where it runs on to the chest. Make your cut leaving plenty of skin attached to the neck, and skin up the neck toward the head. In the case of deer with antlers you must also cut from between the antlers down the back of the neck to get the head skin off. Carefully skin the head, taking care to remove any fat or flesh from the skin as you do so; skin over the head, cutting through the base of the ears, and taking the same care with the eyes, gape, and nose, as previously described with the small mammals. Once you have split the eyes, nose, lips and ears, and satisfied yourself that all fat and tissue have been removed from the skin, set it into a container in which you have previously mixed a phenol solution of one part phenol to eight parts water.

Leaving the skin to pickle in the solution for at least a week, take notes and measured drawings of the skinned skull and neck, then remove the head from the neck at the occiput (where the spinal vertebrae enter the skull). Flesh the skull and scoop out the brains as best you can before putting the skull on to boil. When this is done scrape any stubborn particles of flesh off, until the skull is completely clean.

Now make a baseboard in the same oval shape as the base of the neck, and fix a centre board protruding from it. This centre board should be slightly less in depth than the neck, and you fit your skull

An exceptional roe deer head,
mounted for display

Two recommended cameras: the Minolta X-700 *(left)* and the Canon A-1 *(right)*

A Nikko Stirling scope aligner

A selection of knives recommended by the author: from left to right, Puma bowie with doctored handle; Puma emperor folding lock knife with smooth handle; Westmark; Smith & Wesson general purpose knife; Smith & Wesson skinner; Smith & Wesson lock knife

onto the end of it. There is no easy way around this particular step. It is purely experience that will teach you at what angle to cut your centre board, since it will eventually govern the angle at which the head is held. With the skull firmly in place re-locate the lower jaw into its socket and wire it in place — light fuse wire is ideal for the job. The simplest way to fit the skull to the board is to shape the end of the board so that it fits inside the cranium, and put a screwnail through the top of the skull and into the wood.

When the skull is secure, hammer 1in nails half-way into the wood and 2in apart, along both top and bottom of the centre board and around the outside rim of the baseboard. Now you are ready to start. Make neck muscles using wood wool, shaping them with your hands and binding them firmly with strong, light cord onto the centre board. Anchor the cord to the protruding nails by lacing it from one side to the other. This is not a difficult step, but I advise against too much haste. Make sure the neck is the right shape, thickness etc, using both your measurements and — when nearing completion — the skin, by pulling it over the modelled neck to check for fit. Once again, do not overstuff — the skin may well have stretched when damp and a little shrinkage should be allowed for as it starts to dry.

When you are quite satisfied with the neck, build up the skull by making muscles from gun tow, binding them firmly with waxed linen thread. Eventually you should end up with a fairly exact replica of the head you originally skinned (check your measurements). Fit the skin on periodically to check this, and when you are entirely happy that you have reconstructed the head muscles, hammer all the 1in nails flush, anchoring your binding cord, and smear some modelling clay onto the skull around the eyes and gape. Push some modelling clay inside the ears and fit on the skin, drawing it down around the back of the baseboard. Nail it down around the back, using 1in nails, 1in apart.

Push a sharpened wire, cut to twice the length of the ear, through the tip of the ear and down into the gun tow tied to the skull. Then model the clay inside the ear with your fingers until you are satisfied with the shape. Repeat with the other one. The modelling of the face can be a fairly lengthy process until you become experienced at it. Do not rush the job — if you are getting tired or bored put a damp towel around the head and leave it for a few hours before starting again.

Using modelling clay, place small pieces inside the eye sockets with a modelling tool which you can buy in most art shops, and shape the clay with your fingers. Work down to repeat the same process with the lips and nose. Fit your glass eyes and model the eyebrows and eyelashes. It is very important that you continuously refer to your measurements while mounting the head, and that you are not discouraged if your first attempts don't look professional. You will learn only from experience that as the skin starts to dry, over a period of weeks, if you have not mounted both sides of the head exactly the same the face may twist. It happens to even the best of us. I worked on a polar bear once, and during my absence through 'flu, my colleague modelled the head. It didn't look at all bad until the skin started to dry. As the days went on the polar bear's face began to resemble more and more that of a giant, white bulldog. But we softened the skin down again and re-modelled it — that polar bear is now on exhibition in one of the top natural history museums in the country.

Paint the nose to give it a moist appearance, and cut the protruding ear wires. If you are happy with your head, put it on the wall. If you are unhappy with it have a good look at the area you are not happy with and work out why. The reasons will be either overstuffing, uneven stuffing, or under-stuffing. If you have a ready supply of specimens you can get a fresh one and start again from scratch, or alternatively you can soak the skin of the original specimen, take it off, and do it again.

Skin Dressing

Most sportsmen have at some point regretted their lack of knowledge of skin preparation, particularly when they have a nice, thick winter pelt which they must either throw away or send to a commercial tanner — a disproportionately expensive process unless the skin is intended for clothing. Yet home skin dressing is very simple and can make attractive floor and wall decorations.

Although the equipment needed to dress fox skins to make your wife that special coat is complicated and expensive, and unsuitable for domestic purposes, it is entirely possible to produce quite excellent results. These home-dressed skins should not be used outdoors since they would absorb water, the skin ending up like a chamois leather and, when dry, like a board.

Small Skins — Squirrel to Fox
Skin the animal, cleaning off all fat and tissue. If there is blood or

162

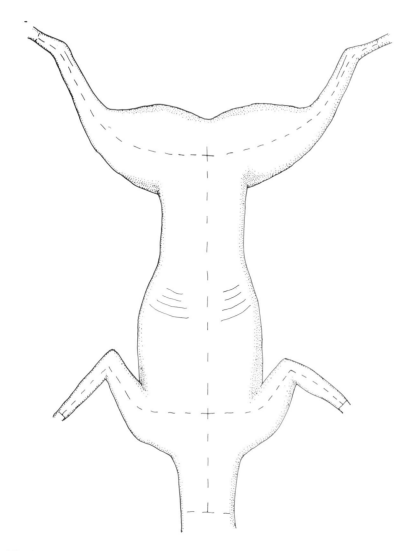

The broken lines indicate where cuts should be made for skinning

soiling on the skin wash it off in lukewarm water, rinsing it thoroughly. Leave it to drain then wash the hair side with any good shampoo and rinse it clean. Rub in some conditioner, exactly as you would do with your own hair, rinsing it clean in running water. Hang the skin up until all excess moisture drains from it, squeezing the skin out as you would do with a piece of material, but do not wring it. With a wide-toothed brush lay the hair in its natural direction to give it a smooth appearance, and place it hair-side down on a wooden

163

board, giving it a gentle drag 1in up the way to flatten any hair which may have become distorted when laid on the wood. Using 1in round nails peg out the key points of neck, or head if still attached, the four legs, belly and tail. Now, gently stretching the skin, carefully work your way around the whole skin putting 1in nails in at 2in intervals (closer for very small skins). With the elasticity of a skin you can continue to go around the skin pulling nails out and stretching it to quite a surprising size, but avoid this or the hair side will start to look thin.

When the skin is well pegged down, and naturally and aesthetically shaped, rub into the skin side a mixture of 50 per cent alum and borax, never salt since it is virtually impossible to remove from the skin tissues once it has been absorbed, and ever after will draw moisture. Rub the alum and borax into all the skin surface and leave it to dry propped vertically to allow moisture to drain.

After a week, when the skin is completely dry, take a flat pumice stone and rub the whole skin in small circular motions to give it a smooth suede-like appearance, dusting it lightly with a little talcum powder, unscented if possible. Then take a sharp scalpel blade and, angling the blade so that the cut on the skin angles towards the edge avoiding a sharp cut in the skin, cut the skin neatly around on the inside of your line of nails, thus avoiding the puckering which the nails will have caused.

When you have taken the skin off the board you will find it is hard and stiff. It may be used as it is for a wall or floor decoration, or for the back of armchairs and so on. If you wish to soften the skin take it between your hands, and with the same motion as used in washing a pair of dirty socks, rub it hard all over to break up the skin fibres, and repeat until you are satisfied. Skin dressed in this fashion can be used quite successfully for anorak collars, knife sheaths, or glove backs. However, if wet, the skin will tend to go hard.

Large Skins
Deer hair is tubular and generally unsuitable as long-term floor decoration and clothing since if it is walked on the hair will break, and bald patches develop. Using the same technique as for a small skin, but omitting the rubbing and softening process, deer skins can be used successfully for decoration throughout the house, where wear is not likely.

Setting Hooves
An attractive gun rack or walking stick handle can be made from the

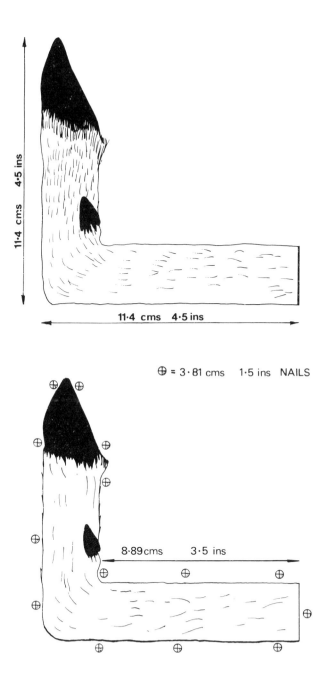

The correct measurements and position of nails for setting a gun rack or walking stick handle

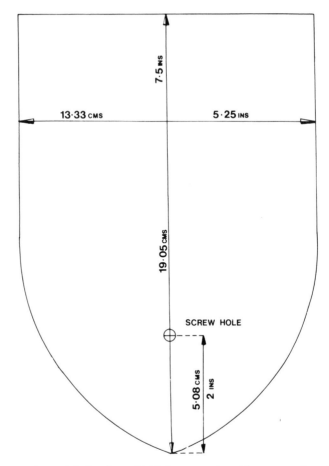

7·5 INS

13·33 CMS   5·25 INS

19·05 CMS

SCREW HOLE

5·08 CMS   2 INS

The correct size and design for a shield, for mounting either gun racks or frontals

hooves of roe deer, and the method of doing this is perfectly simple. Cut your hoof to the measurements given in the diagram and soak it in a phenol solution for a week. Then bend it to exact right angles and set it on a wooden board held in position with 1in nails, and leave it to dry for one week. Then fix it to a shield with a heavy screw nail running through the shield into the bone. The marrow should be hollowed out and filled with hard-setting adhesive or plastic wood when fitting the screw. If the hooves are dull or scored badly a light application of clear nail varnish will give them an attractive and professional finish. For fitting the hoof to a walking stick handle, place a small piece of dowling or metal rod inside the hollowed bone, fixed with adhesive, and fit to the shaft of your chosen stick.

166

Shields

The choice of shield design is entirely personal. However, I advise you to stay away from fancy curved or ornate shapes as they seldom look good or professional, and the design illustrated, in the size suggested, is one which, after much trial, I have found to be most pleasing.

Frontals

For those who do not wish to go either to the expense of having a head mounted or who do not care for these heads protruding from the wall, then the obvious choice is to have a frontal — a much neater and less ostentatious way of displaying antlers. A frontal refers to the area of skull from which the antlers grow.

It is sad that so many otherwise pleasing heads are spoiled and made to look ugly by rough hacking with a saw when a little more care can make a showpiece. The skull should be cut through (see diagram) from the tip of the nasal bone to the occiput, running through the orbital ring (eye socket). Then, after careful boiling, the skull should be bleached, preferably with hydrogen peroxide. If this is not available a mixture of household bleach may be used, though care should be taken and the skull watched. If left too long in this mixture the surface of the bone will erode, giving a granular, unnatural ap-

Skull frontal, showing where to cut the skull for mounting

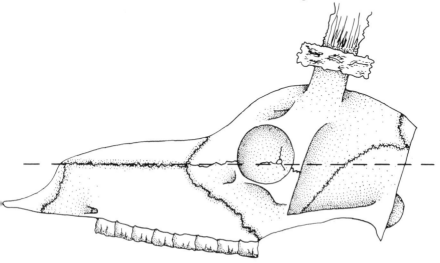

pearance. All contours between the pearling on antlers should be carefully cleaned and scrubbed out, emphasising the detail.

When your trophy is completely dry mix brown oil paint with burnt·sienna, or wood dye of the right colour, and lightly paint the antlers. This is trial and error since you neither want the antlers to be too dull nor too highly glossed, but looking natural. When dry, the points, tines, and highlights may be scratched with broken glass to give a natural, sharpened, white appearance. When mounted on a shield of the design illustrated here the frontal will look entirely professional, natural, and artistic. Again I must emphasise that there is no substitute for practice, but with a little determination and perseverance you can develop a hobby which is highly rewarding.

It is a requirement of the Wildlife and Countryside Act that if you wish to do taxidermy commercially you must be registered. However this does not apply to anyone practising taxidermy for his own pleasure.

Cock pheasant

10 GAME COOKING

Knowing what to do with an animal or bird carcass is obviously very important. These days there are few game purveyors who will dress a bird for you, or butcher a beast. Even if you are lucky enough to live near one, the cost of plucking and drawing a bag of, say, three pheasants can be off-putting, especially when you add it to the day's shooting expenses. It is also correct that the person who has had the pleasure of the day's shooting, whether with rifle or shotgun, should be the same one who has the less pleasant task of preparing the meat for the table. This is something I insist upon with anyone I ever introduce to field sports, whether they are young or old. I believe that anyone who shoots game must be able and prepared to dress the carcass for the oven and then eat it.

All animals and birds on the game list are edible, though some can have a 'stronger' or more 'gamy' taste than others, depending on the time of year, age of the animal, sex, and how long it is left to hang before being cleaned, and so on. The flesh of some birds can have a distinctive and fairly unpleasant taste, because of their diet. In the case of the capercaillie this is almost exclusively the shoots and buds of conifers. Once you know the correct way to pluck or skin your quarry cooking it should present few problems.

There will be few field sportsmen today who do not own a deep freeze, and, indeed, access to one is essential unless you wish to spend the shooting season eating nothing but game and be starved of it for the rest of the year. Since the largest joint or bird you are likely to put away is no bigger than a decent-sized leg of pork, the average family deep freeze should hold all the meat you bring home.

There is another method of preserving meat which, though many hundreds of years old, is now becoming popular again. That is curing and smoking meat, but not using those little home smokers many of us are familiar with. These can be dangerous if not used correctly, increasing the botulisms in the flesh and allowing bacteria to breed inside what is basically a smoky case. Smoking gives a whole new slant to meat (and not just game). The secret of successful smoking is the use of brine solutions which cure the meat before it is smoked. The process is simple — the meat should be steeped in a brine solution for anything from a few hours to over a week, and then popped into a smoker for a few days. The results are interesting and extremely tasty.

However, before you reach that stage you must first prepare the carcass, making it 'oven ready'.

Feathered Game

Hanging

With the exception of waterfowl, all feathered game should be hung for at least 24 hours before being plucked and cleaned. The birds should be hung singly from the neck. Do not be tempted to hang two or three birds together from the same string as air must be able to circulate around them. Hanging tenderises the meat and gives it flavour by allowing decomposition to start. The longer a bird is allowed to hang the more 'gamy' the flavour is. Personal taste will dictate how long you hang your birds. Another factor to take into account is temperature. If it is extremely cold the decomposition process will slow up and will stop altogether if the temperature falls below freezing point. Then you can leave your birds for a considerably longer period.

Although I do not usually leave birds hanging for more than four days, during a particularly cold winter I had a cock pheasant hanging for over four weeks — the carcass was so cold that I could not possibly have plucked it. Conversely if you are going through a warm spell, hang birds for little more than 24-48 hours. Flies may be a nuisance in warmer weather so take extra care with birds, wrapping them in muslin if possible.

If you notice bluebottles' eggs on the bird — they are creamy white and look like minute grains of rice — immediately clean the area on which they have been laid, and when the bird has been completely plucked and singed, wash the flesh thoroughly. Flies prefer to lay their eggs on the moist parts of the bird — around the eyes, in the beak, around the anus, and where the shot has hit. If the eggs have hatched into maggots I do not advise you to keep the bird. Discard it and chastise yourself for wasting good meat unnecessarily. Ducks and geese are not hung as this causes the flesh to rot and the birds become inedible. They should therefore be plucked as soon as possible.

Plucking
Plucking a large bird can be hard work, though it is not difficult. All game birds are plucked in the same manner, and the time taken will largely depend on the size of the bird. Ducks and geese generally take longer since they have an under-layer of down which must be removed. It is a commonly held belief that if you plunge a goose into boiling water, known as 'plotting', it will be easier to pluck. In my experience I have never found it to be the case, dry feathers being considerably easier to remove from goose flesh than wet ones.

Whenever possible pluck geese and ducks outside. Geese in particular, being large birds, give you a greater area, and therefore more feathers, to pluck. Plucking a goose indoors can result in both yourself and your kitchen looking like the inside of a feather pillow, with the downy feathers sticking to everything. Either place the bird on a table or on your lap, with its feet toward you. Pluck the outer feathers a few at a time by tugging them sharply toward the neck of the bird. They should come out easily as the skin is tough. Pluck the wings only as far as the elbow, as there is little flesh beyond the elbow joint. Pluck about three-quarters of the way up the neck and as far as the heel joint on the legs. When the outer feathers have been removed you move on to the down. Lay a damp cloth to one side to moisten your thumb. The best way to remove down is to rub the damp ball of your thumb across the flesh and pick up the matted bits of down. It can be a tiresome task, but you must remove the down before you singe the bird. This is easily done with a gas flame or bottled gas stove. Holding the bird a few inches above the flame singe off the remaining fluff.

Cut the feet at the joint using a sharp knife, bone cutters, or rose secateurs. Remove the tendons in the legs — approximately six in each leg. The head is removed by cutting the neck close to the body. A sharp knife will do the trick.

To draw (or gut) the bird cut the skin vertically from the breastbone to the vent. Reach in with your hand and remove everything you can draw out — the gizzard (second stomach), entrails, gullet, heart, lungs, and windpipe. You may have to pull hard at the windpipe to remove it as it will still be attached to the bird at the neck end. The lungs too can be tricky. They are fastened firmly against the back of the rib cage and the inexperienced can often miss them. They are pinkish and spongy in appearance. Make sure the vent is quite clear, and before rinsing the bird check its crop. If there is anything inside remove the whole crop by carefully edging the crop from the flesh with your fingers and lift it out. For your own education, if the bird has a full crop, open it up. You will be amazed at the amount of grass, seed, berries etc a bird can cram into it.

Now rinse the bird thoroughly inside and out. Some people maintain that rinsing the bird will remove some of its flavour and that wiping the carcass with a damp cloth is sufficient. I disagree. I would rather sacrifice a tiny bit of flavour to make sure that what I was going to eat was as clean as I could make it.

Your goose is now ready for cooking. If you intend to freeze it, let the bird drain, dry it with a clean cloth and place it in a freezer bag. Suck all the air out, seal and freeze the bird. It is inadvisable to keep game, especially waterfowl, for any longer than 18-24 months in the deep freeze, though I have eaten both geese and pheasants that were over three years old and found them to be delicious.

The plucking and cleaning of all feathered game is basically the same as I have described for geese though, as I have already said, you do not have the down to contend with. The easiest bird to pluck is probably the pigeon, its body feathers falling out at the slightest touch. This is because the feathers are loose and the pigeon's skin is strong. Having said that, however, because the pigeon's legs and wings are small they have little flesh, and are really not worth plucking as they will take a disproportionately long time to do. The breast, being plump and meaty is the choicest part of the bird, so it is permissible to remove the skin from the breast and cut out the breast meat as two steaks. This is done by slitting the skin carefully down the breastbone and pulling it back on both sides. The meat is then cut away from the bone with a sharp knife and removed. If you prefer to use the whole pigeon you can pluck and draw it in the same way as you would any other bird. Whichever method you prefer, try to choose a young bird with pinky coloured legs, and allow it to hang for about 24 hours.

The skin of some birds, including pheasants, can be delicate so take care not to pull out too many feathers at one time or you will tear the skin. With all birds the younger ones will be more tender, and

172

more suitable for roasting. If you have brought an old grandad home it will still be edible, though it is better to joint and braise older birds. I remember being invited to a dinner party one evening at the home of an eminent friend who had recently taken up shooting. With great pride he announced that the main dish of the evening was to be a goose which he had shot himself! The assembled company made the correct congratulatory noises and waited in anticipation for the meal. Two hours passed, then three, and still we waited, but the goose was 'not quite tender'. Eventually in desperation our hostess announced that we should eat, and we sat down to a beautifully prepared meal, with stuffed goose drenched in redcurrants as the centre-piece. Unfortunately the goose was as tough as last year's shoes, and virtually inedible. I can only deduce that it was a great grandfather of many years, to have been so tough, as I have never since come across such a bird.

Cooking

A large goose will need to thaw for 24 hours before cooking. As with all fowl make sure any birds are *completely* thawed before putting them in the oven. The process of cooking kills bacteria, but if a bird is cooked while still partially frozen, you risk the bacteria remaining alive, with obvious nasty results to yourself and your family.

I have never considered trussing to be important — it only serves to make the bird look tidy at the table, and makes carving a little easier. You simply bind the wings to the body using either a skewer or a trussing needle and fine string. Push the tail into the slit above the vent and tie the legs to the tail. The skewers are removed immediately before the serving.

There is no mystique in cooking game birds. They require no more preparation than cooking the Sunday chicken, though since most of us eat game only on special occasions it is better to give their preparation a little extra thought. Stuffings for instance are not essential though they certainly enhance any bird. Even if you are not partial to conventional stuffings, placing a peeled apple or two, or even an onion in the body will give the bird extra flavour and moisture.

Recipes

Roast Grouse *(one bird per person)*
Grouse are small birds with a distinctive flavour. Roast grouse are traditionally served on slices of toast which have absorbed the juices during cooking,

with game chips, redcurrant jelly, and bread sauce. You will need the following ingredients per bird:

1 oz butter	1 rasher streaky bacon
1 slice toast	Salt and pepper
Plain flour	Cranberries or redcurrants for stuffing (optional)

Heat oven to 400°F (200°C), gas 6. Insert butter into bird, rubbing it well into the cavity. Stuff with redcurrants or cranberries. Cover breast with bacon rasher. Cover and roast for 20-30 minutes, then remove from oven, drain juices from roasting tray, remove bacon from breast, and drench breast in seasoned flour. Place a slice of toast in roasting tray and place the bird on it. Replace in the oven without covering, and roast for a further 10 minutes until nicely browned. Serve on the same piece of toast, which during cooking will have absorbed the liquid, with gravy from the bird's own juices.

Roast Duck à l'Orange *(serves 4)*
One of the most traditional ways of serving duck, it is still one of my favourites and is hard to beat. Remember that wild ducks are consistently smaller than the domestic variety, and consequently you will need two ducks to serve four people. However, the advantage in using wild stock is that there is little fat on these birds and their flesh is tastier.

2 ducks	1 tbsp sherry or Cointreau (optional)
3 or 4 oranges	½pt stock (if you have kept the
1 tbsp lemon juice	giblets make it from these)
1 tsp tarragon	

Heat oven to 340°F (170°C), gas 3-4. Roast the ducks in a covered dish for approximately 1½ hours. Grate the skin of half the oranges, cutting the remainder of the peel into thin strips. Chop the oranges finely, keeping the juice. Boil up the stock, add the tarragon, the oranges, and the lemon juice. When it is blended together add the sherry or Cointreau. When the duck is nearly cooked remove the cover and allow the skin to brown. Serve it drenched in the stock mixture, with vegetables of your choice, including an orange and cabbage salad.

Pheasant Casserole with Red Wine *(serves 4)*
The time of year a pheasant is shot has a bearing on how much flesh is on it. A bird shot late in the season will be thinner than one shot in the autumn, which is still fat from good summer feeding. Hen birds have a finer textured flesh, and are generally better tasting. One pheasant should serve four people.

174

1 pheasant	2 bay leaves
1oz butter	½tsp rosemary
2 medium onions, chopped	½pt red wine
3 or 4 sticks of celery, sliced	1tbsp brandy (optional)
2 large or 3 small carrots, sliced	salt and pepper
2oz mushrooms, sliced	cornflour

Heat oven to 340°F (170°C), gas 3-4. Joint the pheasant into four pieces. Place the butter in a frying pan, add the onions, celery, and carrots and cook until onions are golden brown. Remove to a casserole. Brown pheasant pieces in the frying pan and place in casserole. Add half of the wine to the pheasant, along with the mushrooms, bay leaves, rosemary and seasoning. Place in the oven, uncovered, for about 1 hour, basting frequently, until cooked. Remove the pheasant from the casserole, add the remainder of the wine and the brandy to the dish, thicken with cornflour and pour over the pheasant. This dish is delicious, especially if served with creamed potatoes.

Roast Goose with Peaches and Cherries *(serves 4-10)*
Most recipe books will tell you that goose is a very oily bird, and therefore heavy to eat. With wild geese, pinkfeet or greylag, this is certainly not the case, and goose remains one of my favourite dishes. Certainly I always prick the skin of the bird thoroughly to allow the fat to drain out, but the flesh is superb, especially when served with fruit. One goose should serve at least four hungry adults, and from personal experience I can tell you that it can stretch to a decent meal for ten.

1 goose	1 tin peach halves
4 rashers bacon	1 tin black cherries
¼-½pt red wine (optional)	Salt and pepper

Heat oven to 350°F (180°C), gas 4. Stuff the bird with any stuffing you like, mixed with plenty of redcurrants or redcurrant jelly. Place the bird in a roasting tray, with rashers covering the breast. Add half of the red wine, cover and roast for 1½ hours. After this time prick the bird to release excess fat, and skim off juice and fat, add cherries and juice, seasoning and peach juice to tray with the remainder of the red wine and return it to the oven for a further ½ hour uncovered. Baste with the juice from the pan. When tender remove from the oven, serve drenched in juice from tray, garnished with peach halves and accompanied by roast potatoes.

Partridge in Vine Leaves *(one bird per person)*
This is a traditional way of serving partridge. Vine leaves are obtainable either fresh or tinned (from delicatessens), or, if completely unavailable, cabbage leaves can be substituted. You will need one partridge per person, ingredients as follows:

| 1 rasher bacon | A little butter, melted |
| 1 vine or cabbage leaf | Salt and pepper |

Heat oven to 350°F (180°C), gas 4. Wrap the bird, which has been seasoned, in the bacon rasher, and cover with a vine or cabbage leaf, tied with string to secure it. Place in a casserole, pour melted butter over bird, add a little more seasoning, cover and cook for about 1 hour, until tender. Remove the vine leaf, serve with apple sauce and game chips.

Pigeon Casserole *(serves 4)*
Since the most fleshy part of the pigeon is the breast, I usually just skin the breast, remove the steaks, and use them for cooking. Although whole roast pigeon is delicious I find casseroled breast steaks are less bother to do and equally tasty.

1oz butter	½pt stock
2 small onions, chopped	¼ tsp marjoram
2 carrots, sliced	Salt and pepper
2oz mushrooms, sliced	A little flour
4 pigeon breasts (8 steaks)	

Heat oven to 350°F (180°C), gas 4. Melt the butter in a frying pan, cook onions, carrots, and mushrooms for 5 minutes to soften. Place in a casserole. Brown pigeon breasts in pan, remove and stir stock in frying pan until boiling. Pour over pigeon in casserole. Add marjoram, salt and pepper and cook in a moderate oven for 1½ hours or until tender, adding a little flour to casserole to thicken gravy. Serve with vegetables of your choice.

Snipe and Woodcock
Snipe and woodcock are two small, delicious game birds. For both you should allow one bird per person, and since they are so small the most effective way of cooking them is by roasting. With snipe you can roast the bird with the head still attached, though the eyes should be removed. Woodcock can be cooked with its innards intact, only the gizzard need be removed, since this imparts a lovely flavour to the bird. The insides are spooned out before serving.

Both these birds (as with the recipe for roast grouse) should be served on pieces of toast.

Furred Game

Rabbit, hare, and deer should all be hung for at least 48 hours to increase flavour and tenderness (see Feathered Game). Seven days should be the maximum hanging time for any furred game. A wrapping of thin muslin will prevent flies from laying eggs on the carcass in warm weather.

Rabbits and Hares

Hares are sometimes hung by the hind legs to allow the blood to collect in the chest cavity, which is used for gravy or in jugged hare. However, if the weather is warm and flies are about I do not bother with this. To skin both rabbits and hares place the animal on a table or draining board of the kitchen sink. With a sharp knife make a slit across the back and ease the skin away from the flesh with both hands. It should come away without difficulty, but if it sticks at all cut the adhering tissue with your knife. Pull the skin right up over the animal's neck and pull through the front legs. Cut the skin away, leaving the animal with two fur socks on its feet. Do the same with the hind legs, pulling the skin over the rear end and pulling the legs through. It will now have four fur socks and its tail attached, with the front half of its skin inside-out over its head. Cut off its head with a sharp knife and using bone cutters or rose secateurs cut off the feet just above the skin line.

To paunch or gut a rabbit or hare lay the skinned carcass on its back and slit the belly carefully to the breastbone. Hold the carcass in your left hand, raising it up slightly and pull out all the entrails. You can pop them directly into a plastic bag for disposal. Make sure the carcass is properly cleaned and using your cutters cut the tail off at the base and open the pelvis to clean its tract completely. Rinse it thoroughly in cold water and allow to drain. Keep the heart, liver and kidneys, as they can be used for stock.

Rabbits and hares are not usually roasted and should therefore be jointed before cooking. The joints are hind legs, forelegs, and saddle. The rib cage can be used for making stock or soup, but does not really have enough flesh to use in a casserole, the many small bones making it undesirable for this use. Follow the contours of the bone and cut off the hind legs at the thigh. Do the same with the forelegs, cutting as close to the rib cage as possible. To remove the saddle cut down the last rib to the back bone and break it by bending it sharply. Cut through any remaining flesh with a sharp knife.

I have found rabbit to be one of the most versatile meats to cook, and probably one of the most underrated. There is a general reluctance to eat rabbit, either because the average family does not like the idea of eating pretty little bunnies or because they are under the impression that all rabbits are infected with myxomatosis. There is really very little one can say on the first reason, and the second too is complete nonsense. Rabbits without signs of myxomatosis are perfectly healthy and make wonderful eating. It is also possible to eat a rabbit with that horrendous disease without suffering any ill-effects yourself, although I would hesitate to recommend it.

When I am asked how to cook rabbit I reply that you can cook a rabbit in various ways. Whatever you can do with a chicken you can do with a rabbit — casserole, roast, pie, fry. My favourite recipe for rabbit is the following:

Southern Fried Rabbit *(serves 4)*

2 rabbits	Breadcrumbs
2 eggs	Fat for deep-frying

Joint the rabbits as above and discard the rib cages (or use for soup). Boil the joints in a little salted water for about 1 hour until tender, and remove from water. Dry the pieces with a clean towel. Beat the eggs and put the breadcrumbs in a bowl. Coat each joint of meat completely in beaten egg, cover in breadcrumbs, and place in deep fat heated to about 400°F (200°C), gas 6 for about 10 minutes until golden brown. Serve hot or cold, with chips or salad.

The average palate will find brown hare more pleasing than the Scottish or blue hare which tends to have a stronger taste. The saddle of the hare can be roasted, and one saddle makes a delicious and satisfying meal for two people. However, the most common method of cooking is stewing or braising, and if cooked in its own blood it is known as jugged hare.

Jugged Hare *(serves 4)*

1 hare (with blood separate)	1pt stock
4 rashers streaky bacon	Salt and pepper
Butter for frying	1/4pt red wine
3 sticks of celery, sliced	2 tbsp redcurrant jelly
1 large or 2 small onions, chopped	Parsley, chopped

Heat oven to 325°F (160°C), gas 3. The hare should be jointed. Fry the bacon in the butter first and then the hare. Remove to a casserole. Cook the sliced celery and onion in the frying pan for about 4 minutes, then add the vegetables to the casserole with the bacon and hare. Add the stock and seasoning to the casserole, cover, and cook in a moderate oven for about 2½ hours or until the meat is tender. Remove the meat from the casserole and add the wine and redcurrant jelly. Allow to simmer for a few moments, taste and add seasoning as required. Then gradually stir in the blood, being careful not to allow it to boil, or it will curdle. Pour this gravy over the hare and serve, garnished with parsley.

Venison

Deer meat, from whatever species, is called venison, and has been the food of the rich and privileged for hundreds of years. Until the nineteenth century the penalties for anyone found with the King's deer were fearful. Today venison is still eaten by only a few, though we are being encouraged to eat more. With red deer farming being tried in many parts of the country it may not be too long before we see cuts of venison being sold in butchers' shops and supermarkets throughout the country, alongside the old favourites of beef, pork, and lamb. Unfortunately red deer farmers must get over consumer reticence to eating venison, since many people believe venison is difficult to cook and has a strong 'gamy' taste. This is not the case, and unless you are unfortunate enough to have been given venison from an old, smelly rutting red deer stag, you should find the meat tender, juicy, and delicious. Roe deer venison is particularly tasty, but because there are many fewer roe than red deer shot, their meat is more scarce and more expensive to buy. A young beast will be more tender and less strong than an older one, and the female of the species is generally better eating than the male. Stags shot during the rut are not recommended unless you like a very strong, 'gamy' flavour to your meat. Continentals are particularly fond of such a taste.

Deer should be hung for about seven days, or longer, depending on the weather, size of animal, and its age. I once left a deer's carcass for three weeks in coolish weather until the surface of the outer flesh looked and felt like old leather. It was without doubt the most tender and delicious venison I have eaten in a long while. I would, however, not recommend anyone to leave a beast for as long as that until he is quite certain he knows what he is doing. It is better to skin a beast before hanging to allow the air to circulate freely around the whole carcass. You must be extra careful, however, to watch for flies if the flesh is completely exposed.

If you have observed the technique of preparatory skinning as laid out in Chapter 7, the beast should arrive at your larder with the skin largely slackened off around most of its body. However, for the purpose of this section on skinning, we will assume that the deer has been delivered to you simply gralloched, still with head and legs attached. It is best to hang the beast from its two hind legs, with the ties around the hooves if they are still attached, and the legs widely separated. If the beast has already been hocked, that is its feet have been removed, then you must tie the rope around the hock.

Neatly cut the skin from the belly incision down the inside of the thigh to the hock. When cutting skin away from a carcass it is always a good idea to cut out the way, ie running the knife with the back of the blade against the flesh under the skin, as opposed to cutting down through the skin, leaving nasty, unsightly cuts into the flesh, which should after all be at all times presented in as neat and professional a manner as possible.

Having cut the skin on the hind legs you next do a similar exercise on the forelegs. You will find that once you have freed the skin from the tissue underneath around the belly cut, that it will peel carefully back, with as little use of the knife as possible, the knife being used only to start cuts. If you peel the skin away from first one hock and around the leg, followed by the other hock, and then cut through the rectal tissue at the anus, the skin will pull off with the greatest of ease, right down to the forelegs. If at any time you find the flesh adhering to the skin, a neat and gentle cut will free the muscle tissue and let the skin come away. In this manner you work the skin down the neck to the back of the head, and cut the head away. You will find the technique of disarticulation quite simple. If you run your knife around the neck where you intend to cut through to the spine, bend the joint forward, and cut on the ball socket through the cartilagenous material, then push the head backwards and repeat the exercise, you will find that with a little twist the head will come away with the skin attached.

Depending on the time of year, the inexperienced skinner may uncover the most alarming of sights — warble fly maggots. Normally they are found on the back. These hideous creatures, however discomforting they are for the deer, do not affect the flesh in any way. They live mainly on red deer, in the area between the flesh and the skin, creating for themselves a little cell, and normally when you pull the skin off a beast a number will be left on the carcass like large, white boils. The best way to remove these is to carefully pluck the membrane between thumb and forefinger close to the creature, and just skin it off. You will find the flesh unblemished.

The heart, liver and kidneys of the deer should not be left to hang with the beast, but eaten as soon as possible. Venison liver pâté is particularly delicious, and one red deer liver is sufficient to make a good quantity of pâté (see recipe).

To joint a deer carcass does take a little practice, but it is something anyone with common sense and a sharp knife can manage with ease.

180

Anyone faced with a large skinned carcass is often daunted by the very size of it. You cannot just chop through joints on a deer as easily as you would a smaller creature. Yet it is simplicity itself if you use a little intelligence. The removal of haunches is a good example. What you are going to do is to cut the haunch off through the ball and socket hip joint and tight up against the pelvis through to the spine above the tail. The whole idea, when fleshing any deer, is to waste as little meat as possible.

If you look at a skinned carcass, whether it is lying in front of you on a bench or hanging up, you will see that it is similar to your own body. Jointing it is not difficult, and the basic cuts are neck, forelegs, saddle, and haunches (hind legs). On smaller deer, roe for example, there is insufficient flesh on the rib cage to make it worthwhile fleshing, and they are better used as a basis for soup. On the other hand, with red deer there is sufficient flesh around the rib cage, once you have removed the forelegs and shoulders, for you either to bone it all carefully or for use as stew or mince, or alternatively for you to cut between each second rib up to the spine and use them as spare ribs or chops. You cannot really go wrong if you stick to this formula. The haunch of a roe deer is too small to start cutting into numerous roasts, whereas the haunch of a red deer is large enough to make several, and it really is a case of cutting it laterally, sawing through the bone to make cuts the size of your choice. However, venison has a high moisture content and there is a great tendency for people to make their joints smaller than they should be, forgetting that shrinkage will occur during cooking.

Venison can be cooked straight from the deep freeze without thawing, but this should only be done in an emergency, for example if you have forgotten to take a joint out of the freezer, as you will have to get the cooking time right, and it gives you less scope for interesting recipes. A red deer haunch, the largest joint you will cook whole, should be thawed for about 24 hours before cooking. Because of the shrinkage factor when cooking venison it is important that your meat is completely sealed, either with cooking foil, or a cold water paste made up with flour and water. It is unnecessary to marinade haunches before roasting, though this can be done easily and will give the meat extra flavour.

The roasting time for venison is the same as for beef — approximately 20 minutes per pound plus 20 minutes over, at 350°F (180°C), gas 4, increasing the temperature for the last 10 minutes to

181

400°F (200°C), gas 6 to brown the outside of the meat. My favourite venison recipe is not for the haunches but for the sirloins — the two strips of meat cut off the top of the saddle.

Venison Sirloins in Red Wine *(serves 4)*

2 sirloin steaks	1 or 2 bay leaves
1oz butter for frying	Salt and pepper
2 small onions, chopped	2 tbsp redcurrant or cranberry jelly
4oz mushrooms, chopped	Parsley, chopped
½pt red wine	

Heat oven to 340°F (170°C), gas 3-4. Brown sirloins in hot butter and place in shallow casserole dish. Fry onions until soft, finely chop mushrooms and add both to casserole. Pour in slightly more than half of the red wine, add bay leaves and seasoning. Seal casserole and place in pre-heated oven for approximately 1 hour. Check meat, adding seasoning and the remainder of the wine if necessary. Remove the sirloins, add the redcurrant or cranberry jelly to the gravy, leave the cover off the dish and allow some of the liquid to evaporate, until it becomes a lovely, thick gravy. Pour over the sirloins, garnish with chopped parsley, and serve with the usual accompaniments.

Venison Liver Pâté

You can use the liver from any variety of deer, it all makes delicious pâté. Obviously red deer liver is considerably larger than roe, and so you will have to tailor the ingredients to suit the size of the liver. I normally use at least two red deer livers at one time to make up a decent batch of pâté, which I then freeze for use at a later date. However, the recipe given below is for one red deer liver made into pâté.

1 onion (optional)	2 eggs
3-4oz butter	Salt and pepper
1 red deer liver	¼pt thick cream
8oz bacon rashers, streaky	1 tbsp brandy
1-2 garlic cloves	butter for sealing

Heat oven to 340-350°F (170-180°C), gas 3-4. Chop and fry onion. Cut liver and about half of the bacon into pieces. Either put this through a mincer or a liquidiser, with the butter. Add the garlic, eggs and seasoning — you will need about 1tsp of salt — add the cream, and mix thoroughly. Line 1lb pudding basins (foil basins are excellent and available from freezer suppliers) with two bacon rashers, cut in half if necessary, and pour liver mixture on top. The number of basins used depends on the quantity of

182

liver. Cover with foil and cook for approximately 1½ hours. When cooked allow to cool and pour melted butter mixed with brandy over the top. Chill or freeze. When serving simply turn out the pâté from the dishes and each mould will be covered with a layer of bacon rashers. Serve with hot buttered toast, garnished with lettuce and lemon.

Red grouse

11 EQUIPMENT

Equipment for the Game Shooter

Unless you are taking part in a formal day's shooting, where you will not have a great distance to walk from the Land-Rover, then the type of stylised up-market shooting clothes normally associated with some of the better London tailors is totally inappropriate, and more suited for impressing other people as to your station than being truly practical. Probably at no time ever in the history of shooting has a sportsman enjoyed such a dazzling array of specialist-designed shooting equipment, some of it obviously developed by extremely practical minds which understand the needs of the field sportsman, while other items, though appearing to be extremely practical, do not live up to expectations when used.

Shooting Clothing
If you are invited to formal shooting, whether it is lowland pheasants or on the grouse moor, then a good tweed suit with knickerbockers, deerstalker or flat cap, hose, and good hill shoes are the required uniform. If on the other hand your preference is for rough shooting or fowling, then your equipment needs to be a great deal more hard-wearing.

Trousers should preferably be knickerbockers, since with their general extra room they make moving, bending and squatting less restrictive. Hose, and for dry weather a good pair of leather, commando-soled boots, are best. The choice of top is really a personal matter, whether it is a tweed jacket, camouflage smock, or padded waistcoat with pockets. The basic requirements are comfort, lack of noise and warmth. For wet weather one of the jacket and trouser

184

combinations made by companies such as Barbour or Peter Storm are most desirable. One tip I would give if you are buying a pair of long, waterproof trousers that you intend to wear with wellingtons, is to cut off the bottom 12in and have your wife turn them in and re-sew them. It is the height of inconvenience when these garments get covered with mud at the bottoms when you are walking through a field; yet shortened they still keep you perfectly dry, while not collecting excess dirt which can be difficult to remove before you get in the car. Wellingtons on the other hand can normally be washed in a convenient puddle, and few men I know actually drive home in them, preferring to keep them inside a plastic bag.

The rough shooter is also going to require a set of thermal underwear for the very coldest weather. There are several makes on the market, but if you can get them, Helly-Hansen, the Norwegian protective clothing specialists, have good warm under-garments, particularly their polar suits. Many of their products are used for work on the North Sea.

Gloves These should be of the fingerless type, preferably woollen, without leather which is difficult to dry overnight. For extremely cold weather I recommend Helly-Hansen work mittens. They have a neat flap across both palms to allow your fingers to slip easily out of the glove when you wish to shoot. My personal choice of gloves in extremely cold weather is sheepskin mittens, with waterproof climbing over-mittens, connected by a light cord which passes up my sleeve, over my shoulders, and down the other sleeve. At a flick of my wrist I can free my hands without fear of dropping my mittens.

Rubber Boots Wellingtons with studded soles are ideal for hard wear and grip, though it has been my experience that they can perish around the ankles if they are wrinkled. Better for general shooting, when you are not expecting to be in deep water, are the short rubber boots with studded soles, commonly sold for wearing on top of fishing waders. I have found these to be ideal, keeping your feet dry without covering your legs.

Leather Boots There is a wide choice of leather boots. My preference is for a pair of high-quality (normally Italian or Austrian) hill boots with commando soles. An alternative is the excellent longer-legged boots made by companies such as Golden Retriever.

Other Equipment

Goose Calls   As a first choice I would pick the Olt, a second choice the Scotch goose call with the black rubber concertina removed.

Duck Calls   First choice is the Scotch call, with the rubber concertina intact. For calling wigeon an ideal call is the whistle from a kettle. With this you can quickly learn to simulate this duck.

Cartridge containers   My preference is for a bag. If you choose a belt, however, I advise you to buy one of the type which contain closed loops, otherwise you will find that, with wear, the cartridges slip right down and become difficult to extract with speed.

Decoy Bag   One of the net haversack variety is ideal, holding up to 24 goose shell decoys, 24 duck, or 36 pigeon decoys.

Decoys   The best goose decoys are the large shell bodies with detachable heads which fit inside each other for easy transportation. The best duck decoys are the full-bodied variety with weighted keels to keep them upright.

Shotguns   I have, in Chapter 1, discussed fairly fully shotguns and their use. The choice of shotgun depends largely on your pocket. My preference is to stick to the Parker-Hale range, as they meet the highest standards of design and manufacture and, being one of the most reputable and long-established British companies, customers are assured of good service, a ready supply of spare parts and, in the event of complaints, access to the Company directly. The problem with some of the cheaper and more obscure makes of guns is that though they may be well advertised with glossy ads, all too often they do not have adequate back-up.

Equipment for the Stalker

Boots   For hill stalking boots should be the very best hill walking boots available, with commando soles and loops for the laces instead of hooks since they can catch in heather and grass roots when you are crawling. Boots for roe stalking and low ground should be soft and light but strong enough to give ankle support. Fell boots are ideal, but the soles must have some form of tread or you will find your footing may be insecure on sloped damp grass.

Socks should be long and made of wool, not nylon, to avoid perspiration, and worn on the outside of trousers and pulled up to the knees.

186

Trousers   Ideally short breeches, finishing just below the knee, but any suitable trouser material may be used. I prefer trousers which my wife has taken in from the knees down, making them tight. This prevents excess material from flapping, and is more comfortable when inside your stockings.

Jackets   The best jacket design is one similar to a parachutist's smock, or anorak that pulls over the head, since when you are crawling you want something smooth down the front that will not open and catch on vegetation.

The most important lesson that any stalker must learn is that it is better to get wet than to wear waterproofs of any nylon or oilskin-type material, since they will only alert the game you are trying to stalk. If you are in any doubt as to a garment's suitability, scratch it with your fingernail, simulating the gentle touch of a twig. If it makes any noise, reject it. Clothing materials should be of wool or cotton and, most important, of muted tones. It is interesting to note that although all the Scottish Highland estates have their own tweed patterns, which at close quarters may look garish, they have been specifically chosen over the years to blend with the natural habitat and colour variations most prevalent in their particular locale.

Hats   A hat is vital since not only does it keep you warm — your greatest heat loss is through your head — but also it serves to break the outline of your head. The peak on deerstalkers can be adjusted to suit the angle of the sun, cutting out glare and reflection on your spectacles.

Rifle Slings   My preference is for a canvas sling on my rifle, and not a more costly leather one for the simple reason that it is much easier to dry if wet, is quieter, and has the added advantage that it can be used as a tie or drag rope in an emergency. Extra bullets are best carried in an oily rag inside a small plastic bag in your pocket. This way they do not rattle.

Drag Ropes   If you are going stalking with an estate stalker then you can expect him to bring a drag rope when ponies or vehicles are not available. If on the other hand you are going alone, then it is important for you to take with you a rope which is light to carry, and strong enough to drag a heavy stag. Most people put the rope over their shoulders when dragging a beast, but I have found it is well worth the trouble to carry with me a short section of stout brush shaft which can also be used as a walking stick. I attach the rope to this, and with my hands behind my back the shaft becomes an extremely comfortable hand-hold when dragging a heavy beast over uneven ground.

For roe stalking it is necessary for most men to take only a coil of 2-3ft of light nylon cord with which to tie the four feet together, then lug the beast over your shoulder.

The sort of detail you must pay attention to is illustrated by this story: I was once stalking behind a friend of mine and kept hearing a tiny 'tap, tap' sound as he was walking. He was carrying an expensive skinning knife in an elaborate sheath, with fold-over button-down top. The 'tap, tap' was caused by the knife rattling inside the sheath. He was not aware of it and didn't think such a tiny noise would make any difference. It would. Yet just a small paper handkerchief pushed inside the sheath silenced the sound.

Telescopes for Rifles   In the cheaper price bracket Nikko Stirling have an excellent range of scopes which are well-made, and offer a good choice of sight window. Of the more expensive scopes Pecar are quite excellent, very well made, and have exceptionally clear optics. As I have already said I do not recommend zoom or vari-power scopes.

Rifle Cartridges   It is vital that you use the same brand of rifle cartridge that you have zeroed your rifle with. For .22s I recommend Winchester Hush-power, or Eley low velocity. They are quiet, perfectly efficient and capable of killing all small game, with the added advantage that they do not make the unnecessary loud noise of the high velocity bullet. For full-bore rifles I have found Norma ammunition to be excellent, extremely accurate, and consistent.

.22 Rifles   There are several high quality .22s on the market — BRNO, Marlin and Beretta. It is important that a .22 should be large enough to feel 'familiar' in the hand, and be as near as possible to the size of your full-bore rifle. The familiarity which you will then enjoy will mean that you will feel so much more confident changing from one to the other.

Full-bore Rifles   A stalking rifle should be a highly efficient tool, a pleasure to use, and attractive. If you choose carefully it should be unnecessary to have to replace it in your lifetime, and taking into account all relevant factors of quality, manufacture, replacement of parts (if necessary), accuracy, and price, I recommend Parker-Hale sporting rifles. I have used a number of different weapons in a variety of stalking situations, and choose Parker-Hale in preference to all others. With the Company's guarantee of accuracy using Norma ammunition, they are for me the natural choice. Other manufacturers of quality are Sako and Mannlicher. However, I feel that they are expensive.

General Equipment

Binoculars

There seems to be a tremendous temptation for many new purchasers to buy the most powerful glasses they can find. The disadvantage of glasses of 10 power and over is that, apart from the great difficulty in holding them still, they are apt to cause headaches and eye-strain if used for spotting, for instance, a large herd of deer or for any prolonged study of birds. I find that 7 x 50, or 8 x 40, are the ideal choices for all outdoor pursuits, giving excellent magnification with a wide field of view. At the same time they have the added advantage of being suitable for continued use.

The modern optics industry, sadly no longer British-based, but Continental, Russian, and Japanese, has brought some really excellent glasses onto the market in recent years, at prices most sportsmen can afford. However, it is important when buying any form of specialist equipment such as binoculars that you should go to specialist dealers for both service and advice.

It is also important when choosing a pair of glasses to follow several simple rules. The glasses must be of a weight and size which is convenient to carry, whether for a day's stalking, or rough shooting when one may have the opportunity to observe something interesting. It is equally important to avoid the tiny pairs of binoculars which, though easy to carry, are unsuitable for all but a quick peep, since being so light, they are difficult to hold steady and focus. Good binoculars should be fairly waterproof, and a handy addition is the detachable, contoured eye pieces which function as blinkers if you are using them with the sun at your back.

Knives

A good knife is probably the most essential piece of a field sportsman's equipment, and yet there seems to be a tendency among the majority of British shooting men to make do with virtually any knife which is attractively packaged and inexpensive. Perhaps it would help if they

189

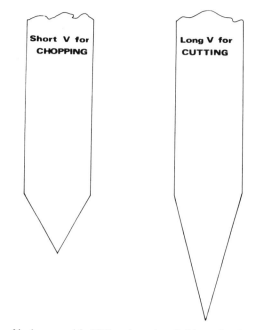

Cutting edges of knives: a wide 'V' for chopping (left), and a sharp 'V' for cutting (right)

understood a little more about the various knife blade designs. Most knife blades are shaped for a specific purpose and all a sportsman needs to do is decide which shape best suits his needs. For general purpose cutting and cleaning the Bowie shape, with a 4-6in blade is best, either as a sheath or folding lock design, depending on personal taste. If used predominantly for skinning a short, swept-back, deeply curved blade gives plenty of scope for long sweeping motions. Avoid the dual-purpose heavy blades with chopping edge on one side and skinning/cutting edge on the other. They are jacks of all trades, and are impractical to use, heavy, and poorly balanced.

With knives, as with almost all other working tools, you get what you pay for. Therefore avoid a cheap knife which will be made of inferior steel and will neither take a fine edge nor hold it when used. In this country there are two ranges readily available which I recommend. The Puma range is excellent, though rather expensive. Their sheaths, however, are something of a let down. Because of the design it is particularly easy to cut the sheath when taking the knife out, but worse, if you fell awkwardly you could easily impale yourself. Smith and Wesson, the prestigious firearm manufacturers, produce a fine range of high quality knives which are available in this country. Each one comes with a secure, professional sheath or belt pouch, and is truly excellent. I would however, suggest that you definitely avoid the cheaper lock-back folding knives, which have an unpleasant tendency to close on your fingers at the most inopportune moments. Some of the Scandinavian knives have no hand guard — decidedly dangerous, particularly if your hands are wet or cold. I have, and use, three different knives, though this is not strictly necessary; a large folding Puma, a Westmark sheath knife, and a Smith and Wesson skinner.

It is very important that a good knife receives good care. Spend time following the sharpening instructions, and if possible use only a Wachita or Arkansas stone if you can get either, as these natural stones cannot be equalled by any of the man-made carborundums. Of course, never throw a knife or misuse a fine edge, as once damaged any knife of hard steel will require long boring work with your stone to replace the edge. A blunt knife is not only useless, but more dangerous than a razor sharp blade which does not require you to hack and saw in order to cut. All knives should always be regarded as extremely sharp, and command respect.

To sharpen a knife, lay your oiled sharpening stone on a flat table or bench and, holding the knife lightly, cut it into the stone toward you, counting the number of strokes, say six, then change the knife to the other hand and cut it toward you another six times. It is paramount that you keep the angle of the blade to the stone consistent. Repeat this until the knife is sharp and can cut the edge of paper. Then reduce the number of strokes systematically, until you are stroking it once per side. Continue this action until the knife can dry shave the hair on your arm. You now have a razor-sharp cutting edge and, unless you misuse it, a few occasional strokes on the oiled stone will keep it keen.

There are two distinct edges one would put on a knife. Imagine the cutting edge as a 'V' (see diagram). A wide 'V' is a strong edge for chopping and rough work and is less likely to be broken or burred, whereas a deep or sharp 'V' is more suitable for skinning and cutting flesh.

Dog training equipment and camera equipment are detailed Chapters 5 and 8.

Ptarmigan

12 NOTES AND RECORDS

Anyone who has any real interest in wildlife, ecology and the out-doors will naturally want to keep an ongoing record, not only of his own progression, but of what he has seen, where, and when. For example, most months of the year I can tell you at what particular times of the day and where you are likely to see roe deer on the ground around where I live. The reason for this is that over the years any time I have seen a roe I have got into the habit of glancing at my watch and making a note of it later in my log. Most of us have seen a standard game book, and they certainly have a value over the years in that you can look back and study when and where you shot a particu-lar species. However, interesting as such game books are, they do not have a fraction of the interest of a proper noted map, and are gener-ally unconstructive and unsavoury, since all they note are what you have killed.

It is never too late to start a proper game map. The easiest way to go about it is to get a large-scale Ordnance Survey map of the area in question or, if that is not available, a map of sufficiently detailed scale, then draw an accurate facsimile of at least 6in to 1 mile, putting in all relevant features such as trees, streams, ponds, etc, numbering them for easy reference.

For your notebook you can use a simple jotting pad; however, I recommend that you go for one of the more expensive bound notebooks, since in years to come you will derive a great deal more pleasure in taking such a professional-looking volume from the book-shelf, whether it is only for your own interest or to show others. The majority of men who start off with a small school jotter, with the in-tention of re-copying it later, seldom do so. It is far better to start off as you mean to go on. Then, simply, as you see a particular species write in your diary 'Roebuck, 9.30am, in field of young barley, 200

192

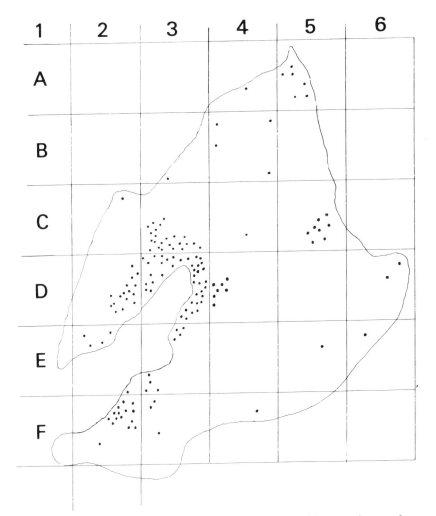

How to mark sightings of individual species on your area, with a sample entry from the diary

4 January 1983	: Red deer stag	C4
7 Janaury	: Hind	D3
8 February	: Hind	E2
16 February	: Group of hinds	B6
1 March	: Stag and hind	D2

yards west of 2'. In this fashion, and by carefully noting any peculiarities of the individual beast, you can, over a period of time, create a diary of great interest in what you have seen, whether it is the fact that you are fortunate to have woodpeckers and you can see that they appear to be multiplying or that at different times of the year different species favour specific areas, normally for food.

193

There is, very near my home, an area where a particular type of ivy grows, along with the weed rosebay willow-herb, and you can see from my notes that in the early spring when the leaves of these plants are at their most tender, roe deer abound there early in the morning. Yet later in the year, as the leaves become tougher and more bitter, you will not find any roe deer there. Then in the depth of winter, when they are really hungry, you will find that they are compelled to return and eat these obviously distasteful plants.

I cannot stress too highly the benefits and pleasure one can get from keeping an accurate record of what you have seen, where and when, since details, if left to even the best memories, tend to become somewhat clouded. As I am writing this I have one of my own diaries beside me, and looking through it I can see that there has obviously been an increase in red squirrels in my area. Another entry recalls when a small, white, tame mouse killed an adder.

Possibly the best example of this technique of record keeping came when some years ago I had a newly-born roe deer orphan, whose mother had been killed by a car. I picked the new-born creature up and took it home, and fed it with a bottle, but after a day or so the little creature would neither eat nor defecate, and appeared to be dying. A telephone call to the zoo gave me no help. I checked through my notebook and could see that I had observed does licking the rectum and genitals of young fawns while feeding them. I propped the by now very sick little fawn up and delicately massaged its hindquarters with my fingers and, hey presto, it defecated. Due to my diary I had inadvertently discovered that very young roe need stimulation to their rectal area in order to defecate. Incidentally that roebuck grew to a most beautiful adult, spending many happy years as one of the star attractions in a deer park.

Do not be shy or reticent about the sort of facts that you include in your diary. Snow is an ideal time to see who has been passing where and when, though I must remind you that an animal has one leg at each corner and can make a lot of prints. If you have studied the tracking section of this book you should find it relatively easy, after a little practice, to be able to distinguish the various animal prints, and do not restrict yourself only to game animals. An example of this is a random entry from my diary for a December day on my land. It reads 'Left the house at 5.45am and drove to shoot. Had hide finished, decoys out, and was sitting by 6.45am, sipping coffee in bitterly cold weather, wondering what the hell I was doing. Around 7.30 I noticed

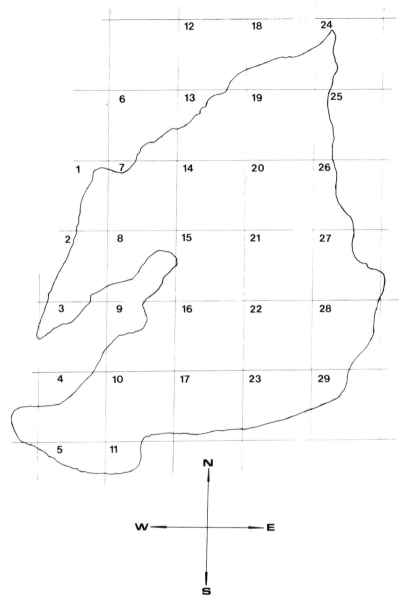

The best and most simple method of making an on-going game map and record, with a sample entry from the diary

3 January 1983 : 7am. Fox; north-west 7; hunting hedgerow
11 January : 6am. Roe buck; south 15; edge of winter wheat
12 January : midday. A lot of rabbits on east 13
1pm. Saw roe doe on east 13
4.30pm. Ducks (mallard) flighting into winter wheat on mid 21

195

a roe doe and two of last year's fawns feeding at the opposite end of the field where I sat. They had appeared as if by magic. Shortly after, the first geese appeared, the roe vanished at the sound of my shots. It was so cold my breath froze on my beard. Sparrowhawk hunting down the hedgerow almost clipped my hat. As the sun came up cock pheasant started to appear, scratching in the snow for food. Two red squirrels were sitting in a rowan bush some 25 yards away, nibbling frozen berries. Large number (approximately twenty) long-tailed tits working up the hedgerow looking for food. Many greylags about with a few pinkfeet, and plenty duck moving around the pond — teal, wigeon, mallard, golden eye and tufties. Returned home mid-morning with four greylags. One right and left, two singles, for five shots fired.' There you are; it took me no more than five minutes to write that, yet all the relevant information is there.

Woodcock

APPENDIX
SHOOTING SEASONS

All dates are inclusive.

*Grouse	12 August – 10 December
*Black grouse	20 August – 10 December
*Ptarmigan	12 August – 10 December (Scotland only)
Snipe	12 August – 31 January
*Partridge	1 September – 1 February
Wild duck and geese	1 September – 31 January (In or over areas below high water mark of ordinary spring tides up to 20 February)
Coot, moorhen & golden plover	1 September – 31 January
*Pheasant	1 October – 1 February
Capercaillie	1 October – 31 January
Woodcock	(Scotland 1 September – 31 January)
*Hares	No close season, but on unfenced non-arable land and moorland may be shot only by the owner or individuals authorised by him, between 11 December (1 July in Scotland) and 31 March. Hares cannot be offered for sale during the months of March to July inclusive.

*In England, Scotland, and Wales it is illegal to shoot these species on Sundays and Christmas Day. The other species may not be shot on Sundays and Christmas Day in Scotland and in certain prescribed areas of England and Wales on Sundays.

Residents of Northern Ireland should check locally for dates of game seasons.

The author recommends that readers contact the RSPB or local police regarding seasons as they are subject to periodic change.

Close Seasons For Deer

All dates are inclusive.

Red	Stags	England and Wales	1 May – 31 July
		Scotland	21 October – 30 June
	Hinds	England and Wales	1 March – 31 October
		Scotland	16 February – 20 October
Sika	Stags	England and Wales	1 May – 31 July
		Scotland	1 May – 31 July
	Hinds	England and Wales	1 March – 31 October
		Scotland	16 February – 20 October
Fallow	Bucks	England and Wales	1 May – 31 July
		Scotland	1 May – 31 July
	Does	England and Wales	1 March – 31 October
		Scotland	16 February – 20 October
Roe	Bucks	England and Wales	1 November – 31 March
		Scotland	21 October – 30 April
	Does	England and Wales	1 March – 31 October
		Scotland	1 March – 20 October

Black cock lekking

FURTHER READING

For anyone wishing to read books of a more specialised nature the author can recommend the following titles:

Chaplin, Raymond E. *Deer* (Blandford Press, 1977)

Douglas, James. *Gundog Training* (David & Charles, 1983)

Hedgecoe, John. *The Art of Colour Photography* (Mitchell Beazley, 1978)

Lawrence, M. J., and Brown, R. W. *Mammals of Great Britain, their Tracks, Trails and Signs* (Blandford Press, 1967)

Ogilvie, M. A. *Wild Geese* (T. & A. D. Poyser, 1978)

Reader's Digest. *Book of British Birds* (Drive Publications, 1969)

Reader's Digest. *Farmhouse Cookery* (Hodder & Stoughton, 1980)

Whitehead, Kenneth G. *Deer of the World* (Constable, 1972)

Whitehead, Kenneth G. *Hunting and Stalking Deer throughout the World* (Batsford, 1982)

Fox

USEFUL ADDRESSES

British Association for Shooting and Conservation (BASC), Marford Mill, Rossett, Clwyd LL12 0HL

British Deer Society, The Mill House, Bishopstrow, Warminster, Wiltshire BA12 9HJ

British Field Sports Society, 59 Kennington Road, London SE1

Game Conservancy Council, Fordingbridge, Hampshire SP6 1EF

The Kennel Club, 1 Clarges Street, London W1Y 8AB

Nature Conservancy Council (NCC), 19-20 Belgrave Square, London SW1X 8PY

Royal Society for the Protection of Birds (RSPB), The Lodge, Sandy, Bedfordshire SG19 2DL

INDEX